Sustainable Civil Infrastructures

Editor-in-Chief

Hany Farouk Shehata, SSIGE, Soil-Interaction Group in Egypt SSIGE, Cairo, Egypt

Advisory Editors

Khalid M. ElZahaby, Housing and Building National Research Center, Giza, Egypt

Dar Hao Chen, Austin, TX, USA

AF864064

Sustainable Civil Infrastructures (SUCI) is a series of peer-reviewed books and proceedings based on the best studies on emerging research from all fields related to sustainable infrastructures and aiming at improving our well-being and day-to-day lives. The infrastructures we are building today will shape our lives tomorrow. The complex and diverse nature of the impacts due to weather extremes on transportation and civil infrastructures can be seen in our roadways, bridges, and buildings. Extreme summer temperatures, droughts, flash floods, and rising numbers of freeze-thaw cycles pose challenges for civil infrastructure and can endanger public safety. We constantly hear how civil infrastructures need constant attention, preservation, and upgrading. Such improvements and developments would obviously benefit from our desired book series that provide sustainable engineering materials and designs. The economic impact is huge and much research has been conducted worldwide. The future holds many opportunities, not only for researchers in a given country, but also for the worldwide field engineers who apply and implement these technologies. We believe that no approach can succeed if it does not unite the efforts of various engineering disciplines from all over the world under one umbrella to offer a beacon of modern solutions to the global infrastructure. Experts from the various engineering disciplines around the globe will participate in this series, including: Geotechnical, Geological, Geoscience, Petroleum, Structural, Transportation, Bridge, Infrastructure, Energy, Architectural, Chemical and Materials, and other related Engineering disciplines.

SUCI series is now indexed in SCOPUS and EI Compendex.

More information about this series at http://www.springer.com/series/15140

Shanzhi Shu · Jinfeng Wang ·
Mena Souliman
Editors

Advances in Geotechnical Engineering & Geoenvironmental Engineering

Proceedings of the 6th GeoChina International Conference on Civil & Transportation Infrastructures: From Engineering to Smart & Green Life Cycle Solutions – Nanchang, China, 2021

Springer

Editors
Shanzhi Shu
Kiewit
Englewood, CO, USA

Jinfeng Wang
Zhejiang University
Hangzhou, China

Mena Souliman
Civil Engineering
The University of Texas at Tyler
Tyler, TX, USA

ISSN 2366-3405 ISSN 2366-3413 (electronic)
Sustainable Civil Infrastructures
ISBN 978-3-030-80141-0 ISBN 978-3-030-80142-7 (eBook)
https://doi.org/10.1007/978-3-030-80142-7

© The Editor(s) (if applicable) and The Author(s), under exclusive license to Springer Nature Switzerland AG 2021
This work is subject to copyright. All rights are solely and exclusively licensed by the Publisher, whether the whole or part of the material is concerned, specifically the rights of translation, reprinting, reuse of illustrations, recitation, broadcasting, reproduction on microfilms or in any other physical way, and transmission or information storage and retrieval, electronic adaptation, computer software, or by similar or dissimilar methodology now known or hereafter developed.
The use of general descriptive names, registered names, trademarks, service marks, etc. in this publication does not imply, even in the absence of a specific statement, that such names are exempt from the relevant protective laws and regulations and therefore free for general use.
The publisher, the authors and the editors are safe to assume that the advice and information in this book are believed to be true and accurate at the date of publication. Neither the publisher nor the authors or the editors give a warranty, expressed or implied, with respect to the material contained herein or for any errors or omissions that may have been made. The publisher remains neutral with regard to jurisdictional claims in published maps and institutional affiliations.

This Springer imprint is published by the registered company Springer Nature Switzerland AG
The registered company address is: Gewerbestrasse 11, 6330 Cham, Switzerland

Introduction

This volume contains eight papers that were accepted and presented at the GeoChina 2021 International Conference on Civil Infrastructures Confronting Severe Weathers and Climate Changes: From Failure to Sustainability, held in Nanchang, China, September 18 to 19, 2021. It contains research data, discussions and conclusions focusing on a number of related geotechnical aspects of infrastructure. Topics include issues related to civil infrastructure such as temperature-induced lateral earth pressure on bridge abutment, subsidence of high-speed rail and expressway, application of recycled rubber mats, railway ballast evaluation, hurricane protection floodwall, tunnel portal stability, deep excavation case study and properties of contaminated soils. Various types of research were used in the various studies, including field measurements, numerical analyses and laboratory measurements. This findings and results should lead to more resilient infrastructure design, maintenance and management. It is anticipated that this volume will support practices regarding the optimal management and maintenance of civil infrastructures to support a more resilient environment for infrastructure users.

Contents

About the Editors

Dr. Shanzhi Shu serves Senior Geotechnical Engineer supporting resolution of geotechnical challenges on Kiewit projects across North America and has been involved in numerous large-scale infrastructure development projects. He is a registered civil engineer in many states of USA and a registered geotechnical engineer in California. He completed his bachelor's degree and master's degree in engineering geology from Hebei Geo University and Jilin University, respectively, China, and doctoral degree in civil engineering from Washington State University in USA. He has about 30 years of both academic and industry experiences. He has authored, co-authored and edited about 30 journal and conference proceedings papers, book and technical reports. He currently also serves as a co-editor of International Journal of Geomechanics (ASCE).

Prof. Jinfeng Wang is Professor of the Department of Civil Engineering, College of Civil Engineering and Architecture of Zhejiang University (ORCID ID: 0000-0002-9099-818X). He completed the Ph.D. on structural engineering at Zhejiang University. He is member of the American Society of Civil Engineers (ASCE) and the Zhejiang Society for Geotechnical Mechanics and Engineering (ZJSGME). As the principal investigator, he has been responsible for over ten significant research projects including the National Natural Science Foundation of China. He has authored, co-authored and edited over 40 of scientific journal papers, books, book chapters and conference papers. He is member of the Editorial Board of Journal of Testing and Evaluation (ASTM).

Mena Souliman, Ph.D., P.E., F. IRF. is an Associate Professor in civil engineering at the University of Texas at Tyler. He received his M.S. and Ph.D. from Arizona State University in civil, environmental and sustainable engineering focusing on pavement engineering. His 12 years of experience are concentrated on pavement materials design, fatigue endurance limit of asphalt mixtures, reclaimed asphalt pavement (RAP) mixtures, aggregate quality, field performance evaluation, maintenance and rehabilitation techniques, pavement management systems, cement-treated bases, statistical analyses, modeling and computer applications in civil engineering.

Dr. Souliman has participated in several state and national projects during his current employment at the University of Texas at Tyler including "Documenting the Impact of Aggregate Quality on Hot Mix Asphalt (HMA) Performance, Texas Department of Transportation" for TxDOT, "Mechanistic and Economic Benefits of Fiber-Reinforced Overlay Asphalt Mixtures" for Forta Corporation as well as "Simplified Approach for Structural Evaluation of Flexible Pavements at the Network Level" which was funded by the US Department of Transportation via Tran-SET University Transportation Center.

Dr. Souliman has also participated in several state and national projects during his employment at Arizona State University and University of Nevada, Reno. He had previously worked as a postdoctoral scholar at University of Nevada, Reno, with the materials and transportation group. He had participated in several national research projects such as the FHWA Project titled "Analysis Procedures for Evaluating Superheavy Load Movement on Flexible Pavements" as well as Asphalt Research Consortium (ARC) projects including "Design System for HMA Containing a High Percentage of RAP Material." Before that, he had worked at Arizona State University where he was the major contributor in the NCHRP 9-44A project entitled "Validating an Endurance Limit for HMA Pavements: Laboratory Experiment and Algorithm Development."

Dr. Souliman has more than 100 technical publications, conference papers and reports in the fields of pavement and aggregate testing, characterization and field monitoring. He is the recipient of the lifetime International Road Federation Fellowship in 2009. In 2017, his research work on pavement engineering-related projects earned recognition as his college's recipient of the Crystal Talon Award, sponsored by the Robert R. Muntz Library, recognizing outstanding scholarship and creativity of faculty from each college as determined by their dean. He also was awarded with the Crystal Quill award in 2018 by the University of Texas at Tyler for his research efforts and achievements.

Response of Approach to Integral Abutment Bridge Under Cyclic Thermal Movement

L. Sigdel(✉), A. AI-Qarawi, C. Leo, S. Liyanapathirana, P. Hu, and V. Doan

School of Engineering, Western Sydney University, Sydney, NSW, Australia
18549889@student.westernsydney.edu.au

Abstract. Integral abutment bridges (IABs) are increasingly preferred over expansion joint bridges due to their ease of construction, and lower maintenance and repair cost. However, no consistent design guidelines exist for IABs due to the complexities associated with soil-structure interactions and nonlinear behaviour of the backfill soil used for the abutments. The temperature-induced cyclic movement of a bridge abutment causes an increase in lateral earth pressure on the abutment wall, which could lead to stress ratcheting. After a number of thermal cyclic movements, permanent soil densification and settlement at the bridge approach will result. In order to study abutment-soil interactions at a different magnitude of cyclic loading, model scale experiments have been conducted, and soil deformations were observed and measured using particle image velocimetry (PIV) technique. Test results show that the number of loading cycles has an important effect on the escalation of lateral earth pressure and soil deformation of the retained soil. Furthermore, the formation of soil settlement and heaving was notably increased with an increase in the amplitude of cyclic movement. Findings of the research presented here assist to define the important parameters to consider in the design of the IABs, besides the possible effects on bridge approach at different magnitudes of cyclic movement.

1 Introduction

Traditionally, bridges are designed and constructed with expansion joints and bearings to accommodate the bridge expansion and contraction due to the temperature changes. However, expansion joints and bearings need regular maintenance as a result of deck movement, traffic loading, trapped debris and moisture intrusion. In this respect, expansion joint degradation is a constant cause of costly maintenance work (Biddle et al. 1997; Wasserman and Walker 1996; White 2007). Failure to maintain the joints would eventually lead to overstressing and structural damage to the bridge structure, which must be avoided. In response to this, Integral Abutment Bridges (IABs) have increasingly become the design method of choice for bridge engineers to eliminate expansion joints and bearings problems. However, there is a complex soil-structure interaction mechanism between the retained soil and bridge abutment during expansion and contraction of the bridge deck due to daily and seasonal temperature changes. The cyclic thermal loading causes an increase in lateral earth pressure on the bridge abutment and abutment foundation (Clayton et al. 2006; England et al. 2000; Kim and Laman 2010;

© The Author(s), under exclusive license to Springer Nature Switzerland AG 2021
S. Shu et al. (Eds.): GeoChina 2021, SUCI, pp. 1–17, 2021.
https://doi.org/10.1007/978-3-030-80142-7_1

Lehane 2011; Springman et al. 1996) and also results in soil deformations adjacent to the abutment.

The pressure exerted by the backfill on the abutment, and extent and depth of the soil settlement behind the abutment are dependent on many factors, including the magnitude and frequency of abutment movement due to thermal effects, properties of the backfill material, and the depth and type of abutment.

Therefore, it is instructive to identify the relative importance of each parameter and its effects on the design of the bridge abutment. Since the increase in earth pressure behind the integral abutments depends on the soil deformations, and the two are mutually dependent, it is necessary to explore the soil-structure interactions associated with the IABs. Many earlier studies (Clayton et al. 2006; Cosgrove and Lehane 2003; England et al. 2000; Kim and Laman 2010; Lehane 2011; Razmi and McCabe 2020; Springman et al. 1996; Sandberg et al. 2020; Tapper & Lehane 2005) had investigated the pressure distribution developed on the abutment due to the expansion and contraction of the superstructure from temperature variations. An experimental study conducted by Huang et al. (2020) to examine the combined effect of abutment-soil interaction and pile-soil interaction under cyclic horizontal displacement has also supports the ratcheting effects behind the abutment. Huang et al. also reported that the shape of the pressure distribution was depend on the magnitude of horizontal displacement.

PD6694–1 (BSI 2011), the current design practice of IABs in the UK incorporated the soil ratcheting effects to establish the soil lateral stress profile during the service period of the bridge. It has specified that for the integral abutments that accommodate the thermal movement by rotation and/or flexure displacement, the design stress profile depends on the at-rest earth pressure condition, the flexibility of the abutment wall, and stiffness of the soil beneath the foundation of the abutment wall. However, for the shorter height end screen abutments such as bank pad abutment and semi integral abutments, which accommodate thermal movement by translational without rotational movement, specified design lateral stress is independent of wall stiffness and soil-structure interaction. The PD6694–1 suggested a linear increase of earth pressure corresponding with the value K_d^* over the top half of the retained height of the wall, but the earth pressure then decreases linearly from $K_d^* \gamma H/2$ to $K_0 \gamma H$towards bottom of the abutment. For abutments subjected to cyclic translational movement, PD6694–1 clause 9.4.4 recommends the following equation to calculate the K_d*:

$$\mathrm{K} * \mathrm{d} = \mathrm{K}_0 + \left(40\mathrm{d}'/\mathrm{H}\right)^{0.4} \mathrm{K}_{p;t} \tag{1}$$

where K_d^*, γ, H, d_d', $K_{p;t}$ are the earth pressure coefficient, unit weight of soil, retained height, wall movement at $H/2$ and passive earth pressure coefficient based on triaxial test, respectively.

As shown in Eq. (1), it is instructive to study the soil-structure interactions under different magnitudes of cyclic thermal movements to establish not only the impact of earth pressure but also the backfill deformation that determine the design of IABs. This study forms part of an investigative effort to establish an engineering approach to either rectify or at least alleviate the build-up of earth pressure and settlement/heave at the approach of an integral abutment bridge. By doing this, the outcome of this research will help to minimise the practical limitations and improve the design of approach embankments to IABs.

In order to capture and understand the complex soil-structure interactions of an IAB with different magnitudes of cyclic translational movements, the particle image velocimetry (PIV) technique was used. PIV is an image-based deformation measurement technique based on digital image correlation principles (DIC), which has been widely used for measuring deformations in geotechnical model tests (Stanier et al. 2015; Deng and Haigh 2018). DIC is an image-based deformation measurement technique that analyses the sequences of digital images captured during tests and calculates fields of incremental displacement at the chosen surface of the test specimens under mechanical loading (Sutton et al. 1983). GeoPIV-RG software is an improved image-based analysis program which implements the principle of PIV using the normalised cross-correlation framework developed by Stanier et al. (2015), and White et al. (2003). The research in this paper used the GeoPIV-RG software implemented in MATLAB to study the soil-abutment interactions. The brief description of the development and performance of the GeoPIV software is given by Stanier et al. (2015), White and Take (2005) and White et al. (2003).

The principle of PIV analysis is graphically presented in Fig. 1. The PIV algorithm tracks the texture (i.e. the spatial variation of brightness) of test patch within the area of interest through a series of images and then produces the trajectories of each test patches in a subsequent image. The initial reference image is divided into a mesh of PIV test patches within the area of interest, and then the highest degree of match between the patch from an initial image 1 (time = t_1) and patch from the following image 2 (time = t_2) gives the displaced position of the same patch (u_2, v_2). The analysis is continued for all the sequential images to obtain the complete trajectories of each test patch.

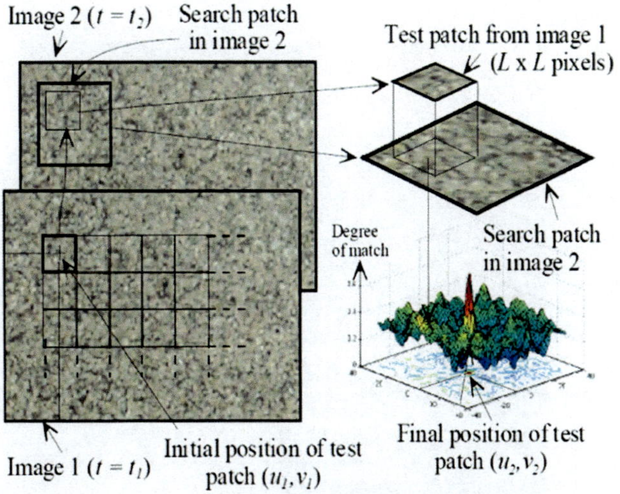

Fig. 1. Overview of PIV-DIC analysis principles (White and Take 2002)

This paper describes the soil structure interactions of an IABs by tracking the soil deformation behind the abutment wall, which is obtained during the cyclic translational movement of an abutment wall within a geotechnical model tank. The cyclic translational movement was imposed on the model wall in an active and passive direction to simulate the bridge deck expansion and contraction during thermal temperature variations. The deformations tracked by the GeoPIV, together with the lateral earth pressure readings are used to study the soil-structure interactions of an IAB. Comparative studies at different magnitudes of wall movements were performed, and the measured pressure distribution and soil deformations are analysed and discussed.

2 Experimental Programme

2.1 Model Tank and Backfill Soil

The laboratory model tank used for the experiment has dimensions of 1000 × 250 × 300 mm (length, breadth, height). It contains a transparent Perspex window of 40 mm thickness at the front and back panels. The model tank contains the retained soil of 660 × 248 × 230 mm (length, breadth, height). A movable vertical wall with dimensions of 300 mm × 248 mm × 13 mm (height, width, thickness) is attached to the tank and connected with a movable bar and two supporting rods to prevent any rotational movement during translational wall movement, as shown in Fig. 2. All three bars are connected to a rigid frame, and the movement of the wall is controlled by an external screw connected to the centre of the rigid frame. The wall was freely sitting on the bottom of the model tank, and the configuration of the rods and wall was set-up in such a way that the movement is genuinely cyclic translational. Eighteen permanent control markers with known object-space coordinates were positioned at the transparent

Fig. 2. The experimental set-up

window to calibrate the image space coordinates to the object space coordinates using the subroutine provided in GeoPIV-RG.

Locally acquired siliceous sand was used as the backfill material. Sieve analysis (Fig. 3) showed that the sand was of uniform size, with particle size ranging from 300 to 425 μm. The sand was loosely placed in the test chamber for all three series of tests, for which the drop height was maintained as 0 mm. For a given value of cyclic translational movement, the integral bridge abutment supporting loosely placed backfill would surely develop the maximum settlement trough; therefore, these experiments limited to the loose backfill. The obtained dry density of the siliceous sand was 1.596 g/cm^3, and the friction angle established from triaxial tests was 30°.

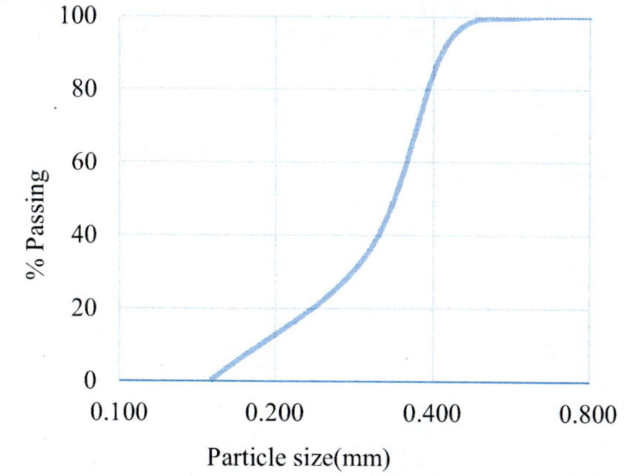

Fig. 3. Particle size distribution

2.2 Instrumentation

Three earth pressure cells of 50 mm outside diameter were used to measure the earth pressure during the tests. The pressure cell was placed at 30 mm, 80 mm and 130 mm height from the bottom of the tank, and all of them were centrally located along the width, and mounted on the thin Perspex sheet, which has a thickness of 10 mm on the side of the backfill soil and attached to the moving wall. A load cell was mounted at 90 mm height from the bottom to measure the lateral force applied to the moving wall. A linear variable differential transformer (LVDT) is installed at the top of the wall to record and monitor the wall movement. The earth pressure cells, load cell and LVDT were connected to the data acquisition unit, and the real-time data were transferred and recorded to the computer using the GDSLAB control and acquisition software.

An 18- megapixel digital camera (Canon EOS 550D) was used to capture the images during the tests for the PIV analysis. The camera was controlled from the computer by the EOS utility software. The camera was positioned at a distance of 500 mm away from the front side of the transparent window for all three experiments, and the same position was maintained throughout the tests. Two white LED lights of 20W used as a lighting source were placed at two sides of the model tank to provide uniform illumination at the exposed face of the model tank.

2.3 Test Procedure

The testing set-up was the same for all three experiments. All tests were performed and analysed for twenty cycles of translational wall movement, with the magnitude of the cyclic translational movement (d) of 0.5 mm, 1 mm and 2 mm, respectively. The wall was initially positioned at the neutral position and moved to the passive direction (inward wall movement) and then active direction (outward wall movement), which represented deck expansion following deck contraction due to the temperature variations. The speed of wall movement was set at 1 mm/min. The negative displacement represents the inward movement of the wall towards the soil, and positive displacement represents an outward movement of the wall away from the soil.

3 Results and Discussion

The formation of the vertical soil settlement and heave behind the wall during the tests was recorded manually in combination with the GeoPIV results. The recorded pressure at three pressure cells, and load cell measurements during the cyclic translational movement of the wall were analysed and interpreted together with the soil deformations observed from the image analysis. For comparison, the recorded lateral earth pressure distribution is presented together with Rankine passive earth pressure and Coulomb passive earth pressure.

The soil deformation after 20 cycles of 0.5 mm cyclic horizontal wall movement is shown in Fig. 4. As we can see from Fig. 4, the soil behind the wall vertically settled down from the original ground surface and somewhere away from the wall, soil started to deposit on the ground surface. The vertical soil settlement and soil upheaving profiles behind the wall after 10 and 20 cycles of translational movement are plotted in Figs. 5 and 6 respectively. The soil profiles in Fig. 5 and 6 show that the amount of soil settlement and soil surface upheaving increases, as expected, with the number and magnitude of cyclic wall movements. The vertical soil settlement after 20 cycles of 0.5 mm cyclic wall movement was 10 mm. At the 20th cycle, the soil settlement has increased by more than 200% for 1 mm of cyclic wall movement, and by 300% for 2 mm of cyclic wall movement in comparison with that for 0.5 mm of cyclic wall movement. The results show that the degree of settlement increases with an increase in the magnitude of the cyclic movement, whereas at a given magnitude, the rate of settlement was at the highest level initially but decreases with the number of cycles.

It was observed that the amount of soil heaving increases with the amplitude of wall movement. The maximum height of the heaving mound is correlated to the movement amplitude. After 20 cycles, the maximum height of the heave was approximately double the height produced after 10 cycles. During $\pm$ 2 mm wall movement, the soil heaved up to a height of 5 mm from the original ground surface after 20 cycles, which is 1.67 and 10 times higher than the maximum heave height obtained for $\pm$ 1 mm and $\pm$ 0.5 mm of wall movement, respectively.

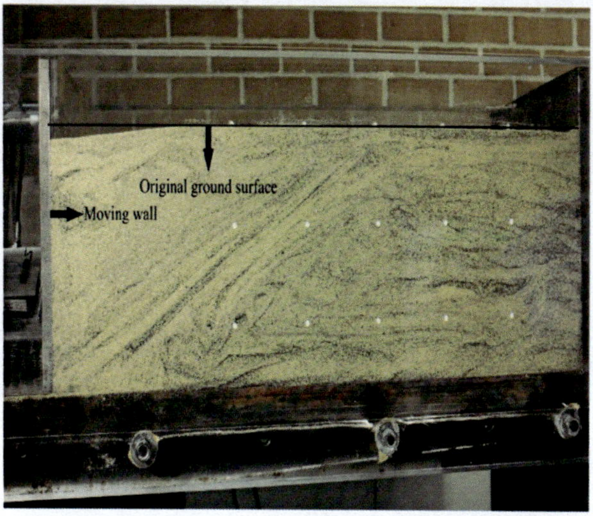

Fig. 4. Soil deformation after 20 cycles of 0.5 mm cyclic translational movement

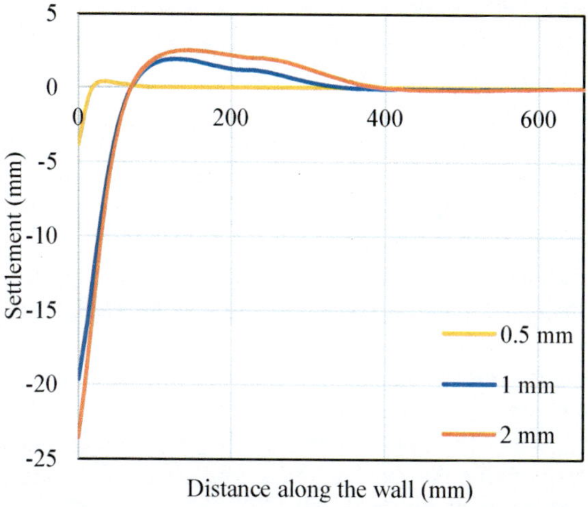

Fig. 5. Settlement profile for three tests after 10 cycles

The calculated resultant displacements of retained soil and vector plots of soil deformation using the GeoPIV-RG are shown in Figs. 7 and 8. The maximum resultant displacement after 20 cycles due to the ± 1 mm of wall movement was 200% greater than that of the ± 0.5 mm of wall movement. The maximum resultant displacement at 20 cycles due to ± 2 mm of wall movement was approximately 12% higher than the maximum displacement obtained during ± 1 mm of wall movement. The presence of the active wedge can be observed for all 3 magnitudes of wall movement after 20 cycles. The width of the wedge was about 60 mm for all three tests from the PIV analyses. The

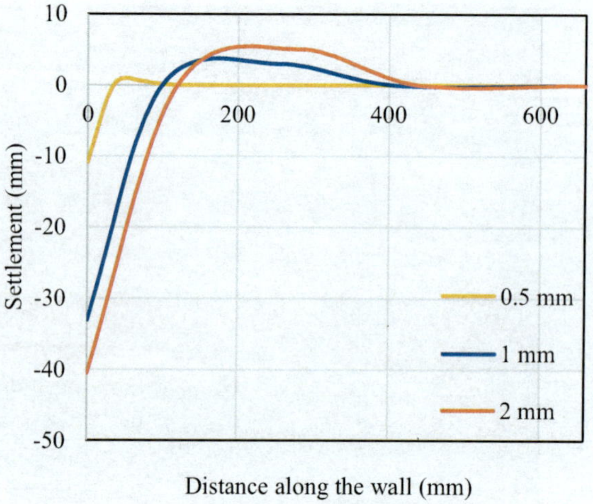

Fig. 6. Settlement profile for three tests after 20 cycles

passive wedge is less well defined and appears at best only at an incipient stage during the ± 1 mm and ± 2 mm of wall movement. Also, as shown in the vector plots in Fig. 8b and 8c, sand was displaced from the toe of the wall towards the surface resulting in surface heave. As discussed earlier, the heaving increased with increasing wall displacement.

The vector plots further confirm that the extent of soil displaced downwards in the proximity of the abutment wall correlates to the magnitude of wall movement. As shown in Fig. 8a, minimal downward soil displacements were observed with the ± 0.5 mm wall movement, but these increased significantly during the ± 1 mm and ± 2 mm of wall movements (Fig. 8b, 8c). Displacements of the sand away from the abutment wall also increased with increasing cycles of wall movement. It is expected that the changes in the density of the sand and arching effects due to particle rearrangements and displacements would result in changes in the lateral earth pressure.

The earth pressure distribution recorded by the pressure cells during the 20th cycle is shown in Fig. 9, where, H is the height of the retained soil and d is cyclic translational movement. The results showed that the earth pressure at the maximum passive position is significantly higher than the recorded pressure at active and neutral positions. The maximum passive earth pressure at the 20th cycle suggests that the build-up of lateral earth pressure was not affected significantly due to changes in the magnitude of wall movement. Here, the maximum passive pressure at the 20th cycle was almost same as that for ± 0.5 mm and ± 1 mm of wall movement. The ± 0.5 mm movement resulted in a maximum passive pressure that is approximately 2.5% higher than that of the ± 2 mm movement. This could be attributed to the soil densification and arching effects behind the wall.

The distribution of the passive earth pressure along the height of the abutment wall after 5, 10 and 20 cycles is presented in Fig. 10. For comparison, theoretical Rankine and Coulomb passive earth pressure distributions are also plotted in Fig. 10. For the

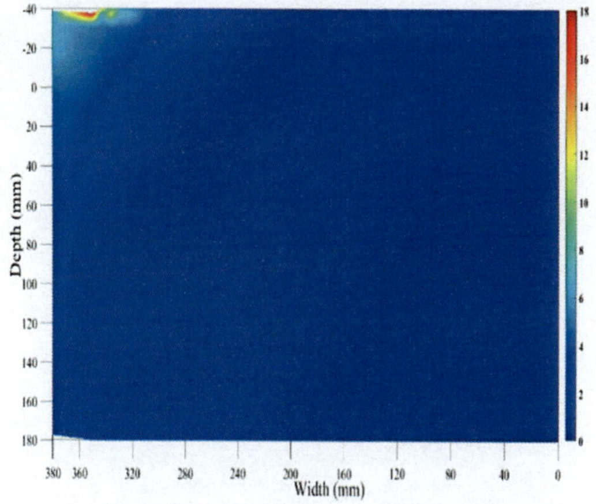

(a)

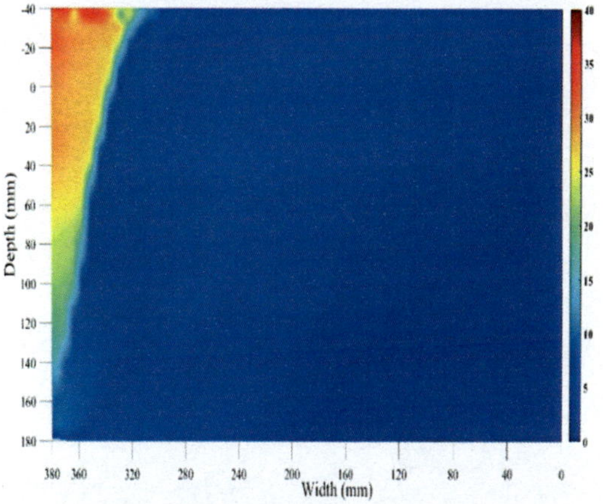

(b)

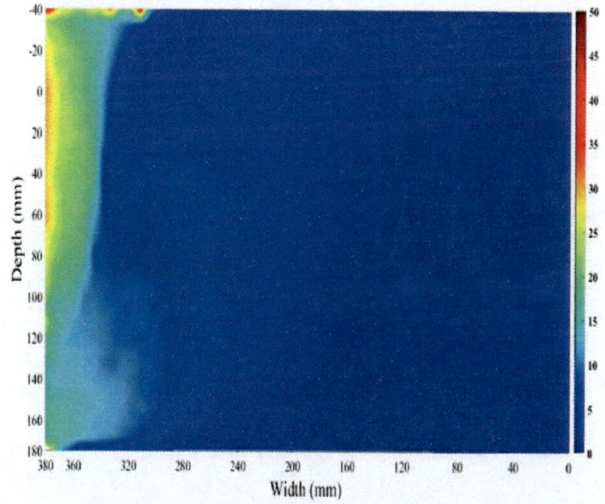

(c)

Fig. 7. Sand resultant displacement after 20 cycles: (a) ± 0.5 mm; (b) ± 1 mm; (c) ± 2 mm

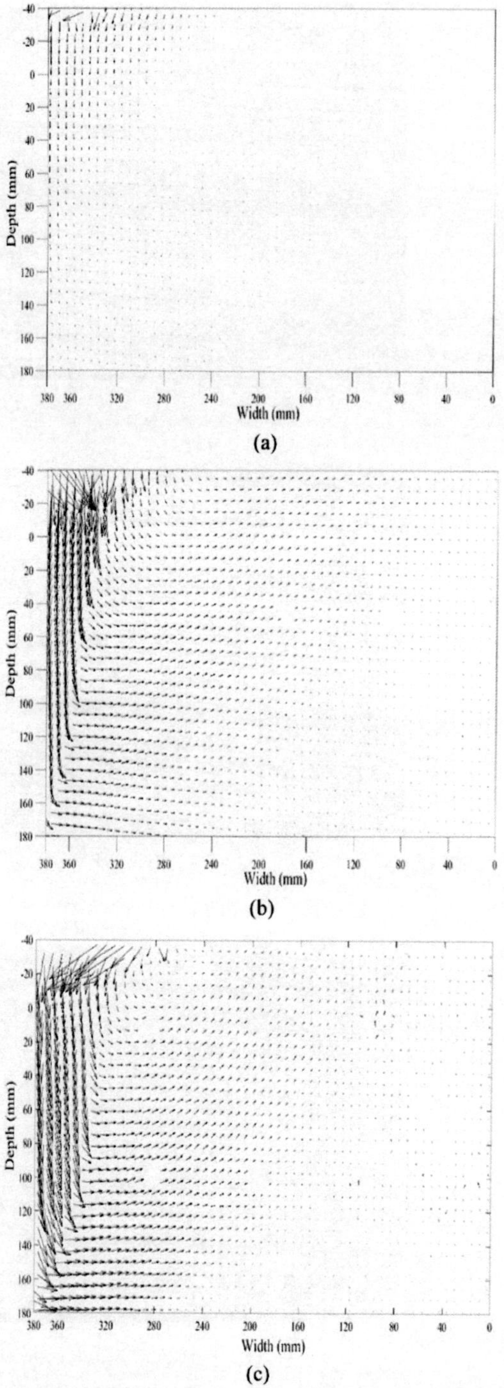

Fig. 8. Vector plots of sand deformation after 20 cycles: (a) ± 0.5 mm; (b) ± 1 mm; (c) ± 2 mm

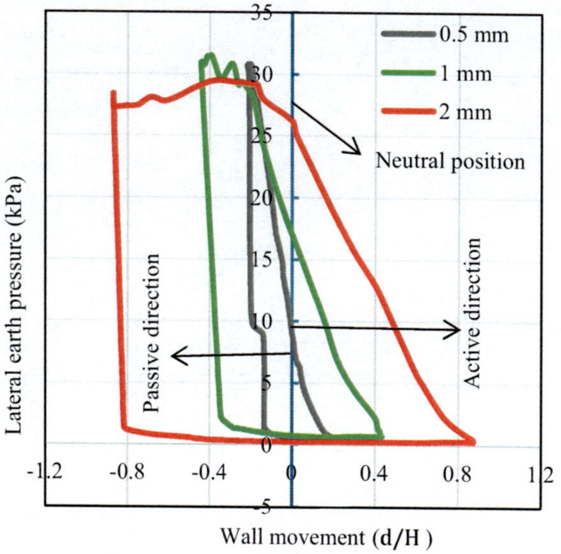

Fig. 9. Lateral earth pressure distribution during the 20th cycle

Coulomb passive earth pressure, soil-wall friction angle (δ) is taken as $\delta = (2/3)\varphi$, where φ is the soil friction angle.

Figure 10 shows that the rate of increase of lateral earth pressure was significantly higher for first 5 cycles, which is approximately 70% of the maximum built-up pressure at the 20th cycle during ± 1 mm and ± 2 mm of wall movement, and 50% of the total build-up pressure during ± 0.5 mm of wall movement. It is also observed that with the increase in the magnitude of the wall movements, the rate of development of the maximum lateral earth pressure also increased.

The passive earth pressure distribution presented in Fig. 10(a) shows that the recorded passive earth pressure eventually exceeded the theoretical Rankine passive earth pressure for all three tests. After five cycles, during ± 1 mm and ± 2 mm of wall movement, the passive earth pressure was greater than the Coulomb passive earth pressure from a depth of 180 mm below the surface to the toe of the wall (at depth 230 mm), whereas, for the ± 0.5 mm of wall movement, the passive pressure was less than the Coulomb passive pressure throughout the total depth of the abutment wall. After 10 cycles (Fig. 10(b)), the passive earth pressure recorded in all three tests exceeded the Coulomb passive earth pressure from a depth of 160 mm to the toe of the wall. In the top half of the abutment wall, the increase of lateral earth pressure is significantly less than the Coulomb passive earth pressure in the bottom half during the 20 cycles of movement. These results indicate that the soil densification, yielding and arching effects of a quite complicated form led to the increase in lateral earth pressure in the bottom and lower middle region of the abutment wall (up to 100 mm from the toe).

The maximum passive pressure with number of cycles at different height location of the abutment wall is plotted together with the Rankine and Coulomb passive earth pressures in Figs. 11, 12, and 13. Is has been shown that the rate of increase of the passive

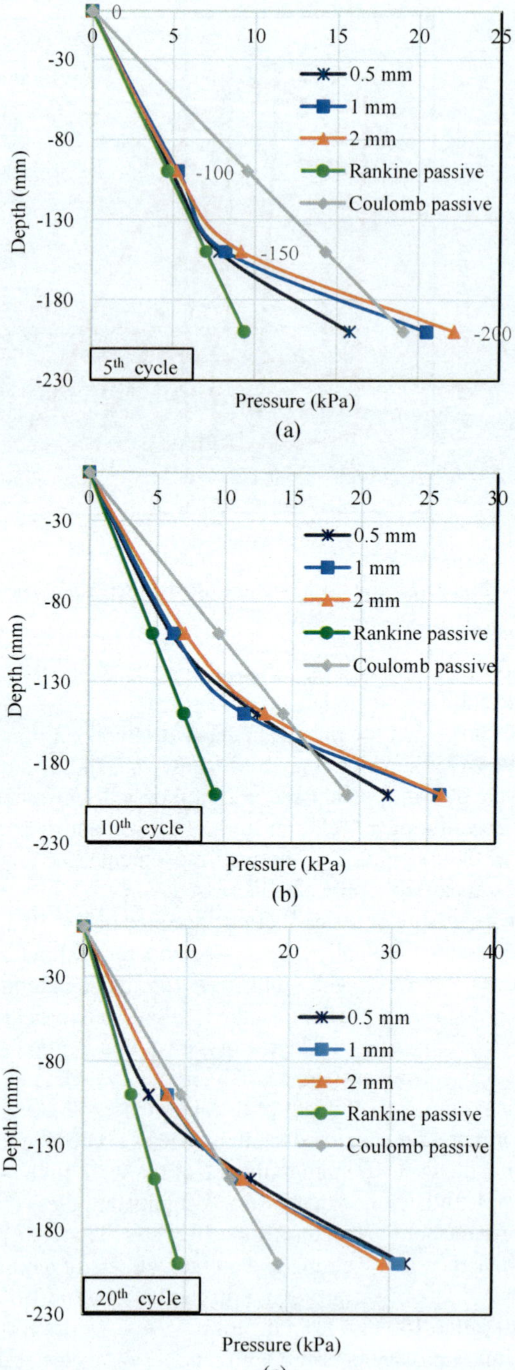

Fig. 10. Distribution of passive pressure along the height of the wall after (a) 5 cycles; (b) 10 cycles; (c) 20 cycles

pressure at pressure cell 2, which is located at a height of 80 mm from the bottom, is higher than the lateral earth pressure recorded at pressure cell 1, and pressure cell 3. These results show that the pressure is continuously building up at a relatively higher rate in the middle region than the bottom region of the abutment wall. As mentioned above, this indicates that continuous particle displacements and rearrangements (see Fig. 8) leading to soil densification/loosening, yielding and arching effects helped to increase the soil pressure in the middle region of the abutment wall.

The earth pressure at pressure cell 1 remained below the limit of Coulomb passive earth pressure for all three experiments. However, it exceeded the Coulomb pressure after between 13 to 17 cycles for all three tests at pressure cell 2, and after 8 cycles or less for all three tests at pressure cell 3. This implies that passive lateral earth pressure prediction using the Coulomb and Rankine passive theories may under predict the earth pressure of integral bridge abutment subjected to thermal temperature variations.

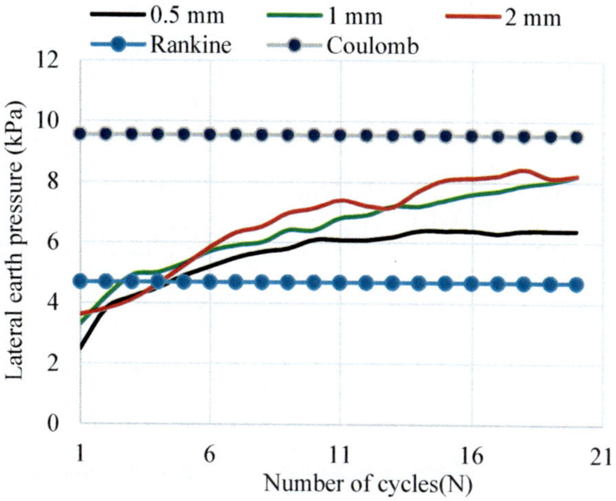

Fig. 11. Passive earth pressure variation with the number of cycles at pressure cell 1 (130 mm from bottom)

The maximum passive force recorded at each cycle of all three experiments is shown in Fig. 14. It shows that the passive force is increased with the increase in the magnitude of cyclic translational movement; however, only a 10% increase of maximum passive force is recorded with 100% increase in the magnitude of wall movement. The rate of increase of passive force is observed to decrease with increasing number of cycles until gradually levelling off as the maximum residual force is reached, which is in agreement with the variations of lateral earth pressure with increasing number of cycles.

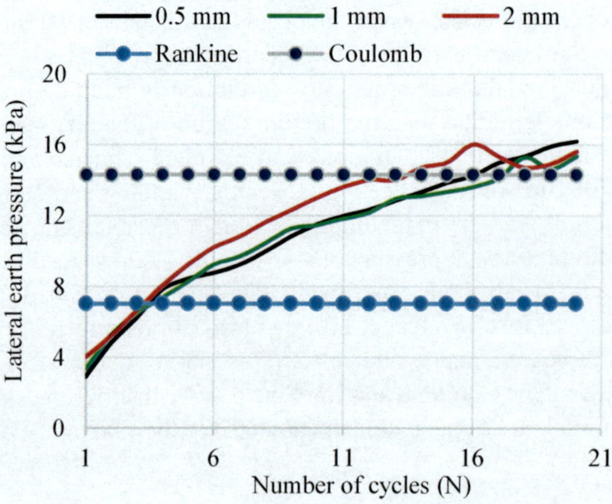

Fig. 12. Passive earth pressure variation with the number of cycles at pressure cell 2 (80 mm from bottom)

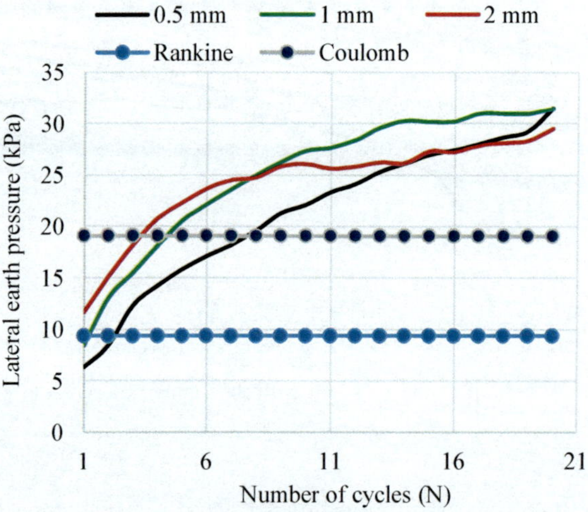

Fig. 13. Passive earth pressure variation with the number of cycles at pressure cell 3 (30 mm from bottom)

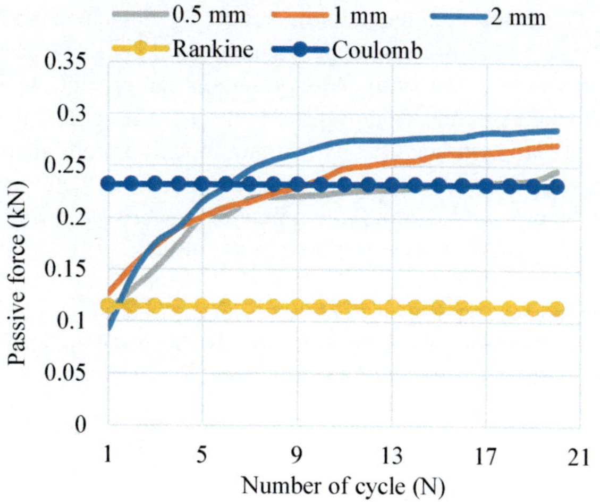

Fig. 14. Comparison of maximum horizontal passive force with the number of cycles

4 Conclusion

This paper discussed the interactions of an integral abutment wall and the retained soil performed at three different magnitudes of cyclic translational movements (0.5 mm, 1 mm and 2 mm) in a laboratory model tank. Experimental observations and GeoPIV analysis results is used to study the stress variation together with the soil deformation during the wall movement. Main conclusions based on the analysis results are outlined below:

1. Thermal-induced cyclic translational movement of wall causes the continuous soil settlement and heaving behind the abutment wall. This increases with the increase in the magnitude of wall movement and the number of cycles of wall movement.
2. With an increasing number of cycles and magnitude of wall movement, soil densification/loosening, passive yielding and arching effects started to develop and an incipient mobilised passive wedge-shaped region started to form. The kinematics is quite complicated. Therefore, it is necessary to develop a better understanding of the design lateral earth pressure profile which takes into account the influence of density, yielding and arching effects in the backfill soil as a result of complex soil-structure interactions, which is still lacking in current design practices.
3. The pressure variations along the height of the abutment show that the maximum passive pressure resides at the bottom of the abutment. This distribution and GeoPIV analysis suggest that the higher lateral stress increases correspond to the locations where the most soil densification and yielding occurred. Therefore, it seems that the stress profile does not correspond to that suggested by PD6694–1, where the maximum lateral stress is located at mid-depth of the abutment. The results suggest that further clarification may be needed for the abutments with shorter height subjected to cyclic translational movements only.

In reality, the seasonal temperature variations will superimpose with the daily temperature during the seasonal changes, which could generate stress profiles different to what was observed in this study. Also, based on the type and flexibility of the abutment wall, and boundary condition at top and bottom of the abutment wall, the movement of the abutment wall could be translational, rotational, or combination of both. The study of the soil-structure interactions for different modes of wall movement, including the erratic movements of the wall, which may characterise the yearly, seasonal and daily temperature variations will be explored in the future.

Acknowledgements. The authors would like to acknowledge the technical supports from the technical staffs at the Western Sydney University for their continuous assistance during the experiments and to Western Sydney University Graduate student research fund.

References

Biddle, A., Iles, D.C., Yandzio, E.: Integral Steel Bridges: Design Guidance, SCI Publication (1997)

Clayton, C.R.I., Xu, M., Bloodworth, A.: A laboratory study of the development of earth pressure behind integral bridge abutments. Géotechnique **56**(8), 561–571 (2006). https://doi.org/10.1680/geot.2006.56.8.561

Cosgrove, E.F., Lehane, B.M.: Cyclic loading of loose backfill placed adjacent to integral bridge abutments. Int. J. Phys. Model. Geotech. **3**(3), 09–16 (2003)

Deng, C., Haigh, S.: Soil movement mobilised with retaining wall rotation in loose sand. Phys. Model. Geotech. **2**, 1427–1432 (2018)

England, G.L., Tsang, N.C., Bush, D.I.: Integral bridges: a fundamental approach to the time–temperature loading problem, Thomas Telford (2000)

Huang, F., Shan, Y., Chen, G., Lin, Y., Tabatabai, H., Briseghella, B.: Experiment on interaction of abutment, steel H-Pile and soil in integral abutment jonitless bridges (IAJBs) under low-cycle pseudo-static displacement loads. Appl. Sci. **10**(4), 1358 (2020)

Kim, W., Laman, J.A.: Integral abutment bridge response under thermal loading. Eng. Struct. **32**(6), 1495–1508 (2010). https://doi.org/10.1016/j.engstruct.2010.01.004

Lehane, B.M.: Lateral soil stiffness adjacent to deep integral bridge abutments. Géotechnique **61**(7), 593–603 (2011). https://doi.org/10.1680/geot.9.P.135

PD6694-1: Recommendations for the design of structures subject to traffic loading to BS EN 1997-1:2004+A1. BSI (British Standard International), London (2011)

Razmi, J., McCabe, M.: Analytical and computational modeling of integral abutment bridges foundation movement due to seasonal temperature variations. Int. J. Geomech. **20**(3), 04019189 (2020)

Sandberg, J., Magnino, L., Nowak, P., Wiechecki, M., Thusyanthan, I.: The integral bridge design concept for the third runway at Heathrow, UK, Thomas Telford Ltd., vol. 173, no. 2, pp. 112–120 (2020)

Springman, S., Norrish, A., Ng, C.: Cyclic loading of sand behind integral bridge abutments, TRL REPORT 146 (1996)

Stanier, S.A., Blaber, J., Take, W.A., White, D.: Improved image-based deformation measurement for geotechnical applications. Can. Geotech. J. **53**(5), 727–739 (2015)

Sutton, M.A., Wolters, W., Peters, W., Ranson, W., McNeill, S.: Determination of displacements using an improved digital correlation method. Image Vis. Comput. **1**(3), 133–139 (1983). https://doi.org/10.1016/0262-8856(83)90064-1

Tapper, L., Lehane, B.: Lateral Stress Development On Integral Bridge Abutments, pp. 1069–1075. CRC Press/Balkema, Boca Raton (2005)

Wasserman, E.P., Walker, J.H.: Integral abutments for steel bridges (1996)

White, D., Take, W.: Particle Image Velocimetry (PIV) software for use in geotechnical testing. University of Cambridge, Department of Engineering (2002)

White, D., Take, W.: Discussion on application of particle image velocimetry (PIV) in centrifuge testing of uniform clay. Int. J. Phys. Model. Geotech. **5**(4), 27–31 (2005)

White, D., Take, W., Bolton, M.: Soil deformation measurement using particle image velocimetry (PIV) and photogrammetry. Géotechnique **53**(7), 619–631 (2003)

White, H.: Integral abutment bridges: Comparison of current practice between European countries and the United States of America, Transportation Research and Development Bureau (2007)

Check for updates

Numerical Assessment on the Influence of Various Factors for Subsidence at the Intersection of Expwy 78 and High Speed Rail of Taiwan

Muhsiung Chang[1](✉), Ren-Chung Huang[1], Chih-Ming Liao[1], Togani C. Upomo[2], and Rini Kusumawardani[3]

[1] Department of Civil and Construction Engineering, National Yunlin University of Science and Technology (YunTech), Yunlin County, Douliu City, Taiwan
changmh@yuntech.edu.tw
[2] Graduate School of Engineering Science and Technology, National Yunlin University of Science and Technology (YunTech), Yunlin County, Douliu City, Taiwan
[3] Department of Civil Engineering, Universitas Negeri Semarang (UNNES), Central Java, Indonesia

Abstract. Choshui River alluvial fan-delta (CRAFD) has been and still is the single-largest subsiding area in Taiwan. Some key infrastructures, including Taiwan High Speed Rail (THSR) and Expressway 78 (Expwy78) that come across this area, are suffering serious problems by the subsidence. Due to complexity of the issue, causes of the subsidence and their influences are not easily identified and quantified. This paper tends to disclose the above puzzles through a numerical approach that would allow separate applications of various factors and examination on their individual effects. The intersection of THSR and Expwy 78 is our primary concern since the subsidence at this location has become most serious along the entire route of THSR and the induced differential settlements are threatening the safety of the transportation artery. An 8-year subsidence monitoring data at the study site is used for calibration of material data adopted in the subsequent analyses. Results indicate the computed subsidence is 147.0 cm, or a rate of 10.53 cm/year, for a period approximately starting with the Expwy78 construction (01/1998–12/2011) and covering complete ranges of construction of Expwy78 and THSR of the site. Contributions due to various factors would be the greatest by previous overpumping (55.2%), followed by soil creeping (14.1%), groundwater fluctuations (11.8%), Expwy78 loading (11.6%), and the least by THSR loading (7.3%). Assessments also indicate around 1/5 and 4/5 of the total subsidence, respectively, occurred as the compression in soils with depths <70 m and >70 m, where 70 m is the average installation depth of THSR piles. Compression at shallower depths would be the greatest by Expwy 78 loading, and thus may trigger a major part of negative skin frictions and discount bearing capacity of the piles. Compression of deeper soils would be caused primarily by the previous overpumping, and would hence lead to significant settlements and distortions of vertical alignment of THSR structures.

© The Author(s), under exclusive license to Springer Nature Switzerland AG 2021
S. Shu et al. (Eds.): GeoChina 2021, SUCI, pp. 18–36, 2021.
https://doi.org/10.1007/978-3-030-80142-7_2

1 Subsidence Issue

Land subsidence has been and is currently as well a significant threat to the living environment and transportation facilities to the western alluvial plains of Taiwan. Figure 1 indicates the locations of previous and present subsiding areas in the island. Since 1950s, Taipei basin has experienced substantial subsidence due to the excessive extraction of groundwaters for municipal purposes (Chen et al. 2007). Started approximately from 1970s as the booming of fishery farming (Fig. 2(a)), the west-coastal plains of Taiwan have become subsided as a result of overpumping of fresh groundwater, which also caused serious seawater intrusions into the ground (Hung et al. 2010). The groundwater tables were estimated lowered by more than 20 m along the coastline and about 4 m to the eastern boundaries of the plains (Chia et al. 1996, Chen et al. 2010).

As seen in Fig. 1, the western plains are still subsiding at present, especially in the area, Choshui River alluvial fan-delta (or CRAFD), with a measured maximum subsiding rate of 6.5 cm/year in 2019 (WRA 2019). The CRAFD is the largest alluvial deposit in Taiwan, encompassing Changhua County to the north and Yunlin County to the south. The deposit was formed by the longest river (Choshui River; Fig. 1) of the island and its tributaries, as well as by the influences of the sea (Taiwan Strait) (CGS 1999). In the past 200,000 years or so, the sea levels has been raised and lowered by $\pm$ 100 m several times in the history due to glacial retreats and advances (Chappell and Shackleton 1986), and thus contributed to the marine (delta) sedimentations to the western portion of the CRAFD.

In recent decades, previously significant subsidence areas along the coastline have gradually moved inlands. Figure 3 shows the shift of the subsidence hot spots in Yunlin County, the southern part of CRAFD, during the period of 1992–2011 (WRA 2011). The shift in subsidence areas was due mainly to the economic growth in the mid-CRAFD, as well as influences arise from the construction of several key transportation arteries, such as Taiwan High Speed Rail (THSR) and Expressway 78 (Expwy 78), passing across the area (Fig. 3). Excessive pumping of groundwaters for the purposes of agricultural, industrial and municipal activities has caused significant subsidence in the mid of CRAFD and threatened the safety of these transportation facilities.

The transportation safety of THSR has become a public concern since early 2000s. The alarm finally triggered in 2009 as the differential settlements between THSR piers have become so great that approached the design limits in angular distortion of 1/1000 for simple supported viaducts and 1/1500 for continuous viaducts (Hung et al. 2010, WRA 2011). The most serious section of THSR that suffered vertical alignment distortions was located at its intersection with Expwy 78. Since the construction period of Expwy 78 was approximately parallel to that of the previously-said subsidence shifting, one would suspect that, besides excessive pumping, the Expwy 78 embankment loading could also have had influences on the differential settlements of THSR piers.

Figure 4 is a photo taken at the intersection area, indicating Expwy 78 embankment as well as THSR piers that suffered the most angular distortions. Table 1 presents the subsidence (i.e., pier settlement) history at the intersection of Expwy 78 and THSR. Although the subsidence rate appeared decreasing in 2000s, the magnitude of THSR pier settlements remained significant and was detrimental to the transportation safety of the facility. Figure 2(b) shows an agricultural pumping well and the vegetation field

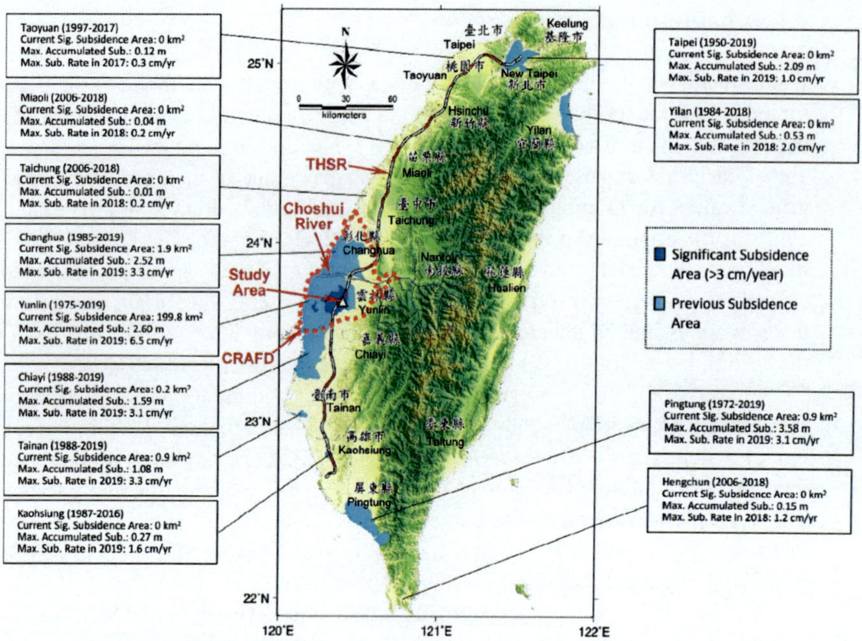

Fig. 1. Ground subsidence regions in Taiwan and study area (base map: WRA 2019; https://www.wra.gov.tw/cp.aspx?n=3679).

(a) Fishery farming (b) Agricultural planting

Fig. 2. Fishery and agricultural farming along west-coastal plains of Taiwan extracting excessive groundwaters and causing ground subsidence.

adjacent to the THSR, suggesting the groundwater extraction for agriculture purposes would be influential to the THSR safety as well.

It is obvious that the factors affecting on the subsidence of intersection area and the transportation safety of THSR would have included the excessive pumping of groundwater and the loading of Expwy 78 embankment. In fact, the loading of THSR piers as well as the groundwater fluctuations and the natural creeping of soil deposits would also be the potential causes for the subsidence. The normal consolidation of soil deposits due to natural deposition process, however, is generally believed to have completed since the CRAFD deposits where the project site is located was formed in late Quaternary

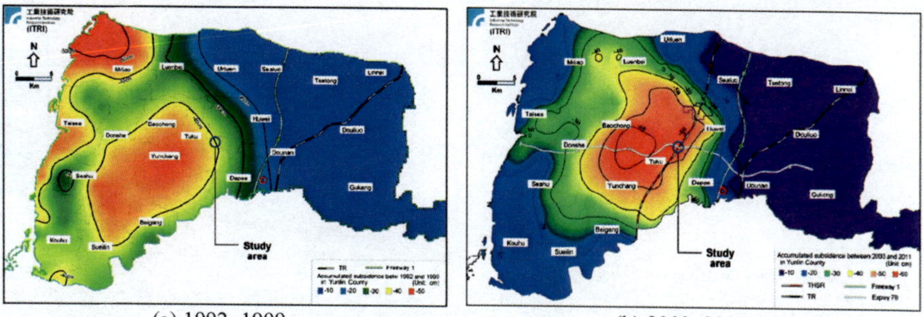

(a) 1992~1999 (b) 2003~2011

Fig. 3. Transition of subsidence areas in Yunlin County, the southern part of CRAFD, between 1992 and 2011 (base map: WRA 2011).

Table 1. Yearly subsidence at the intersection of THSR and Expwy 78 (WRA 2011)

Year	Yearly subsidence (cm/year)	Accumulated subsidence (cm)
2004	15.0	15.0
2005	10.6	25.6
2006	8.7	34.3
2007	7.5	41.8
2008	7.0	48.8
2009	6.5	55.3
2010	6.8	62.1

Note: Average subsidence rate = 8.87 cm/year

period in about 200,000 years ago (CGS 1999). In view of enormous thickness of soils in CRAFD (>750 m; CGS 1999, Hung et al. 2010), we anticipate a small amount of secondary compression (or creeping) of the soil deposits would be existing even if the normal consolidation of deposits due to natural deposition process could have completed for quite a long time.

The aim of this study is therefore set to determine the influences of various factors on the subsidence at the intersection area of THSR and Expwy 78, with particular interest in the estimation of their potential effects on the compression of soils at separate depth intervals of 0–70 m and >70 m. It is noted that the length of THSR group piles in the intersection area is around 70 m below the ground, as shown in Fig. 5. Land subsidence or soil compressions would have effects on the THSR pier structure in two folds. For the deposit with depth >70 m, the compression in soils would settle THSR pier structures and hence distort the vertical alignment of THSR route. Alternatively, for the deposit with depth <70 m, the compression in soils would cause negative skin frictions on THSR piers and thus detriment to the bearing capacity of the group piles.

Fig. 4. Intersection of THSR and Expwy 78 Embankment (viewing NE).

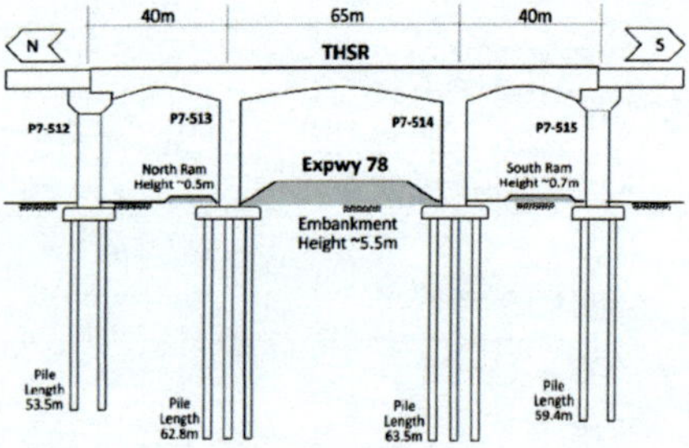

Fig. 5. Schematic illustration of THSR group piles and Expwy 78 embankment.

A numerical tool is adopted in this study for the assessment of influences of various factors on the subsidence at intersection area and the compression of soils at different depth intervals due to various factors. As shown in Fig. 6, three computation points (A, B and C) on the ground surface are assigned. Point A is located approximately 200 m away from both THSR and Expwy 78 and the subsidence at this location is considered with negligible influence by the loadings of these two transportation facilities. Point B is located along the THSR route approximately 200 m south of the intersection. The subsidence at this location is considered to be affected primarily by the THSR loading, in addition to the subsidence at Point A. Point C is located at the intersection of THSR and Expwy 78. The subsidence at this location is due mainly to the loadings of these two transportation facilities, in addition to the subsidence at Point A.

By assuming applicability of the principle of superposition, the influence of THSR pier loading can be assessed by subtracting the computed subsidence at Point A from that at Point B. Likewise, the influence of Expwy 78 embankment loading can be evaluated by subtracting the computed subsidence at Point B from that at Point C. The influences of soil creeping, excessive groundwater pumping (or termed as "previous overpumping" in the following sections) and groundwater fluctuation can be estimated in the numerical computations at Point A by switching on or off separately during the application of previous overpumping or groundwater level variations. Results of the numerical assessment will be discussed in Sects. 3 and 4.

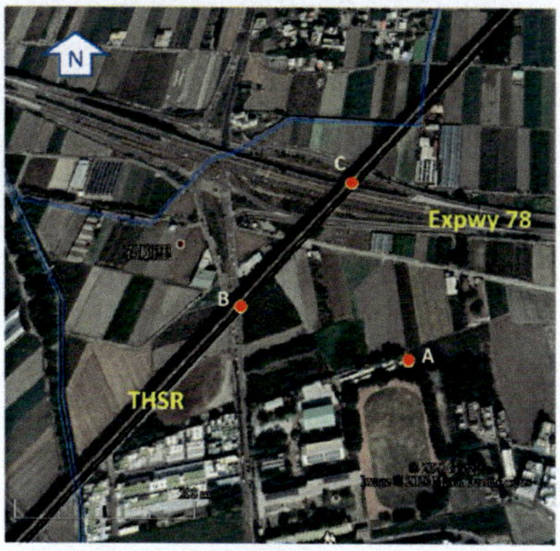

Fig. 6. Location of study area and calculation points in numerical simulation.

2 Background Information

Prior to numerical simulations, some background information pertaining to natural or human factors that would have had impacts on the subsidence of the study area is described herein. These factors include site geology and hydrogeology setups, groundwater overpumping history, and THSR and Expwy 78 construction histories at the site.

2.1 Geology and Hydrogeology

As mentioned previously, the CRAFD was formed by alluvial depositions of Choshui River and its tributaries as well as by marine depositions due to seawater level changes in glacial periods some 200,000-year ago. Hence, the CRAFD consists of alternating layers of alluvial and marine deposits (Chiang et al. 1996, CGS 1999).

As shown in Fig. 7, a typical EW section across the fan-delta and the study area, the alluvial deposits mainly contain coarser particles like cobbles, gravels and coarse-medium sands, with their sizes distributed becoming finer from east to west. The alluvial deposits are generally abundant in groundwater and thus the water resources (aquifers) of the area. Alternatively, the marine deposits include finer particles, such as fine sands, silts and clays, which are distributed along the western coastline and near-shore regions. As the characteristics of low hydraulic conductivity, the marine deposits become barriers (aquitards) in the CRAFD. It is noted that the marine deposits to the western coastline or near-shore regions are generally unified and become a closed boundary of the CRAFD, thus render the aquifers in CRAFD as an underground reservoir (CGS 1999).

The thickness of soil deposits in CRAFD is still uncertain, however, limited documents indicated that no bedrock was encountered up to a depth of 750m below the ground surface (Hung et al. 2010). Within the upper 300m deep, four sets of aquifer-aquitard layers are identified, where the second aquifer (Aquifer 2 or F2; Fig. 7) has been the major water resource layer of the area (CGS 1999). In recent decades, however, the depth for groundwater pumping has become deeper and reached the third or fourth aquifers (F3 or F4) due to the demands of economic growth in the mid of CRAFD (Hung et al. 2010).

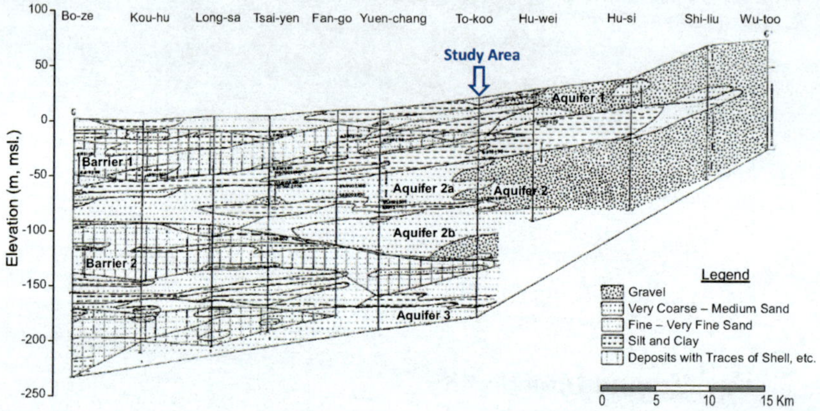

Fig. 7. Geology and hydrogeology profile of the area (based map: Chiang et al. 1996).

Figure 8 illustrates the observed piezometric levels in Aquifers F1, F2 and F3 at some monitoring stations near the study area. As seen, the piezometric levels in F2 and F3 are substantially lower than the one for F1, indicating excessive groundwater pumping has been occurring for some time in these layers and their influences are still continuing. We notice that the long-term trend of piezometric levels in F2 and F3 remains slightly declining, suggesting the influences of previous overpumping in deeper aquifers are not appreciably easing or recovering.

2.2 Previous Overpumping

Substantial lowering of piezometer levels in deeper aquifers of CRAFD as mentioned above was due mainly to excessive extraction of the groundwaters, and its influence on the subsidence of the site had been enormous. As shown in Fig. 3, the inland movement of subsidence hot spots of Yunlin County, in a period between 1992 and 2011, was in relation to the economic growth of mid-CRAFD, that involved more groundwater consumptions from deeper ground for agricultural, industrial and municipal activities.

It's difficult, however, to clearly define the period of excessive groundwater extraction of the area. In view of the observed inland shifting of subsidence hot spots that started

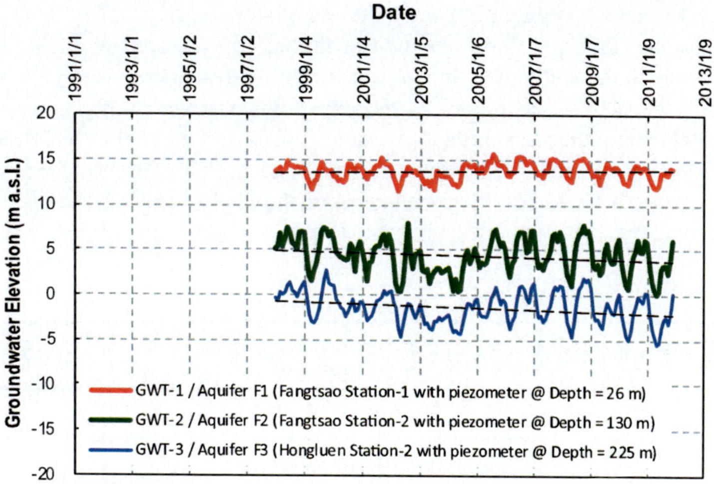

Fig. 8. Groundwater levels of confined aquifer layers in CRAFD (Chang et al. 2020).

approximately in mid 1990s (Fig. 3(a); Hung et al. 2010, WRA 2011, Li 2013), as well as a new agriculture polity promulgated in about the same period that encouraged local farmers to raise more water-consumptive vegetables instead of rice (Ho and Wang 2012), we notice that the significant drops in groundwater levels of the study area might have occurred approximately in the period of 1995–1997, and then became more or less stabilized ever since. This period, from the beginning of 1995 to the end of 1997, is therefore termed as the "previous overpumping" stage and adopted in the subsequent numerical simulations.

2.3 Expwy 78 and THSR Construction Histories

In the study area where two major transportation arteries meet, the loadings of these facilities (THSR viaduct/pier structure and Expwy 78 embankment) would play important roles on the land subsidence as well as the interactions on either of the facilities.

As mentioned previously and shown in Fig. 5, the THSR facility consists of viaducts and piers (i.e., group concrete piles; a pile arrangement of either 3 × 4 for P7–513 & P7–514 or 2 × 3 for other nearby group piles) with each pile a diameter of 2 m and an installation depth of around 70 m into the ground. Although point loads are adopted in the numerical simulations as the way of application of the loading by THSR facility, it should be noted that, since loading is applied on the pile head, the loading actually transmitted through a considerably rigid member of the pile along its shaft and at the base into the ground. Hence, THSR loading on the ground would involve influences in both shallower and deeper deposits.

For the loading by Expwy 78 embankment, a strip-type of surface loading is assumed in the numerical simulations. The loading of Expwy 78 embankment would primarily affect the compression of soils in shallower depths.

The other issue regarding the influence of subsidence behavior would be the timing of loading. Figure 9 illustrates the timeline of constructions of THSR and Expwy 78 facilities in the study area. In contrast to the short-period construction of THSR piers and viaducts, the construction of Expwy 78 embankment at the intersection had taken a much longer time, involving three major stages encompassing the THSR construction of the site. In view of the characteristics and timing of the loading, we expect the subsidence behavior as well as influences on the THSR facility of the area would be more complicated due to the involvement of lengthy construction of Expwy 78 embankment.

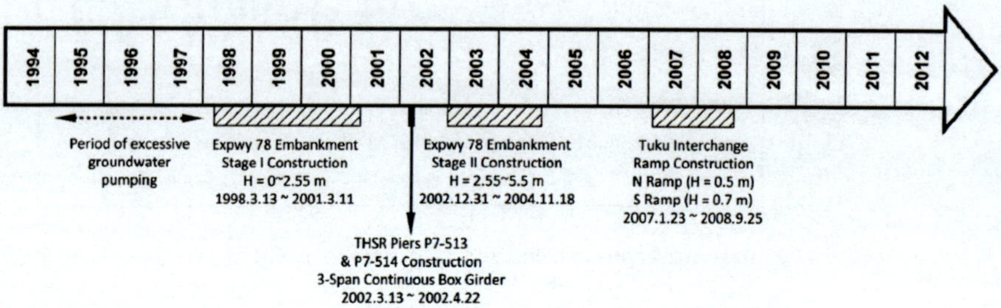

Fig. 9. Timeline for THSR and Expwy 78 constructions.

3 Numerical Simulation Scheme

A numerical software, Plaxis3D, is employed for the assessment of subsidence of the study area and influences due to various factors. As shown in Fig. 10, a 3D numerical model is set up which mimics the onsite arrangement of the two major transportation facilities (Fig. 6; THSR and Expwy 78). Material stratification of the site is assumed in terms of aquifers and aquitards revealed by the subsurface explorations with boreholes to a depth of 300 m from the ground. The loading and timing of constructions of THSR viaducts/piers and Expwy 78 embankment are based on the associated construction drawings as well as the timelines indicated in Fig. 9. As shown in the construction reports of THSR, the vertical loadings on P7–513 and P7–514 are approximately 38,500kN each, while the loadings for the remaining piers of the study area are in the range of 25,000–26,000kN (Sinotech 2003). As illustrated in Figs. 5 and 9, the width of the Expwy 78 embankment is about 55 m and the final height of embankment is 5.5 m in the intersection area. Significant drops in groundwater (or piezometer) levels and subsequent groundwater fluctuations are based on the previous overpumping period as mentioned above and the observed groundwater fluctuations (Fig. 8) at the nearby stations of the study area.

Details of the numerical simulations can be referring to Chang et al. (2020). Figure 10 indicates the numerical model adopted in the analysis, including 4 sets of aquifer (red color)/aquitard (light blue color) to a depth of 300 m, based on results of subsurface exploration of the site, and a 300 m-thick soil layer, termed as Base Soil (light yellow color), to represent remaining soil deposits in the deeper ground. The figure also shows the locations of THSR piles, represented as point loads acting on pile heads, and Expwy 78 embankment, represented as a strip loading on the ground. Generally, numerical simulations are conducted preliminarily with an assumed material data set to verify the results of 8-year onsite monitoring at Points A ~ C (10/2003–12/2011; Fig. 6). The assumed material data is calibrated and modified such that the computed subsidence of the ground at Points A ~ C and compressions of various material layers are comparable with those of the onsite monitoring. Results indicate the computed subsidence at Points A, B and C are 54.6 cm, 59.1 cm and 69.4 cm, respectively, which compare well with the monitoring data of 55.7 cm, 61.2 cm and 70.6 cm. The numerical simulations are then employed to predict the subsidence behavior of the site with a much longer computation period (01/1998–12/2011) which covers full-ranges of construction of THSR and Expwy 78 facilities at the site. With this longer period of computation, the influences of THSR and Expwy 78 loadings can therefore be completely assessed.

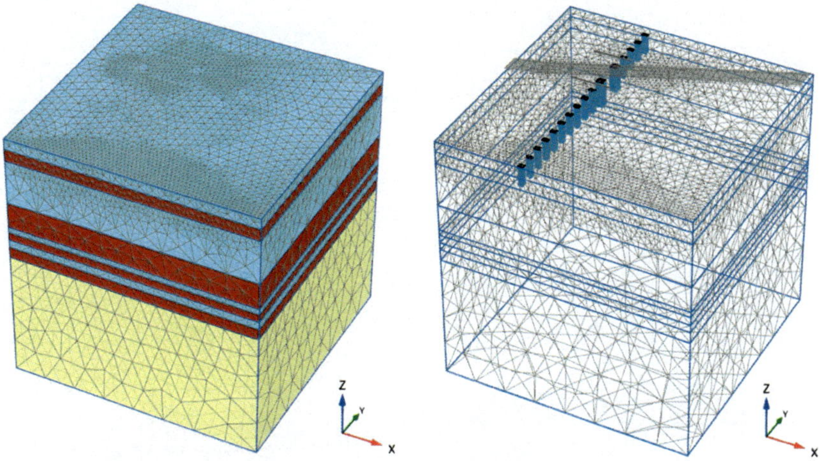

(a) Material layer stratification (b) FEM meshes, Expwy 78 and THSR loadings

Fig. 10. FEM model adopted in numerical simulations.

4 Numerical Assessment Results

The aims of numerical simulations in this study are to examine the contributions of various factors on the subsidence of intersection area of THSR and Expwy 78, as well as the influences of various factors on the compression of soils with depths <70 m and >70 m, that would have dissimilar effects on the safety of THSR viaduct/pier structures. Results of the numerical simulations are discussed as follows.

4.1 Contributions of Various Factors on Ground Subsidence

Results of numerical simulations on the subsidence of study area due to various factors are shown in Fig. 11 and summarized in Table 2. The computation period started from

1998/1/1 and ended on 2011/12/31, a 14-year time which covers full-ranges of THSR and Expwy 78 constructions of the site. As indicated, the computed total subsidence at the intersection area reaches 147.0 cm, or an average subsiding rate of 10.53 cm/year. This subsidence has included the influences due to groundwater-related factors, such as previous overpumping and soil creeping, as well as by the loadings of THSR and Expwy 78 facilities.

Table 2. Computed subsidence due to various factors at the intersection of THSR and Expwy 78 in a duration between 01/1998 and 12/2011

Influence factor	Subsidence of the ground		Compression in soils 0–70 m deep		Compression in soils >70 m deep	
	(cm)	(%)	(cm)	(%)	(cm)	(%)
Soil creeping	20.68	14.1	0.85	2.7	19.83	17.2
Previous overpumping	81.25	55.2	8.91	28.3	72.34	62.6
Expwy78 Embankment loading	17.09	11.6	13.51	42.9	3.58	3.1
THSR pile loading	10.67	7.3	5.55	17.6	5.12	4.4
Groundwater fluctuation	17.35	11.8	2.66	8.5	14.69	12.7
Sum:	147.04 (10.53 cm/year)	100.0	31.48 (2.25 cm/year)	100.0	115.56 (8.28 cm/year)	100.0

Figure 11 visualizes the influences of various factors on the subsidence of the site. It's obvious that the previous overpumping contributes the greatest (55.2%) to the subsidence, followed by soil creeping (14.1%), groundwater fluctuation (11.8%), Expwy 78 loading (11.6%), and the least by THSR loading (7.3%). It is noted that over 80% of the subsidence is due to groundwater-related factors (i.e., soil creeping, previous overpumping, groundwater fluctuation), and the contributions by loadings of THSR and Expwy 78 account for less than 20%.

It is also noted that the computed compression in soils with depths <70 m is 31.5 cm, or 21.4% of the total subsidence (note: average compression rate of 2.25 cm/year). The computed compression in soils with depths >70 m would be 115.6 cm, or 78.6% of the total subsidence (note: average compression rate of 8.28 cm/year). Apparently, around 1/5 of the subsidence occurs in shallower depths (<70 m) of the deposit and would tend to develop negative skin frictions and thus potentially impair the bearing capacity of THSR piles. On the other hand, around 4/5 of the subsidence would be generated at deeper depths (>70 m), which would settle the THSR piles and hence distort the vertical alignment of THSR route.

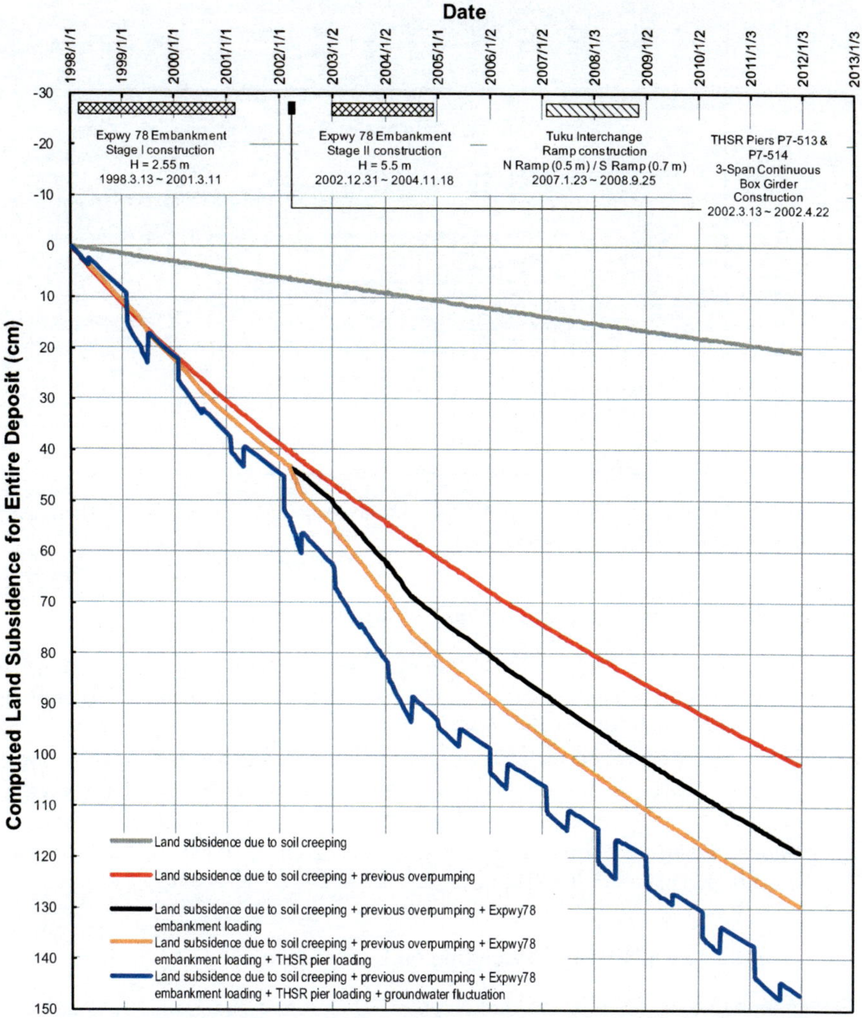

Fig. 11. Influence of various factors on subsidence at the intersection of THSR and Expwy 78 between 01/1998 and 12/2011.

4.2 Influence of Soil Creeping on Compression of Soils

Figure 12 shows the influences of soil creeping on the ground subsidence as well as the compression of soils with depths < 70 m. In view of significant thickness of the deposit (600 m thick in the numerical model), a majority portion (19.83 cm) of subsidence due to soil creeping occurs at greater depths (>70 m), only a small amount (0.85 cm) would generate at shallower depths (<70 m). Hence, soil creeping would be more influential to the distortion of THSR vertical alignment than the bearing capacity of THSR piles.

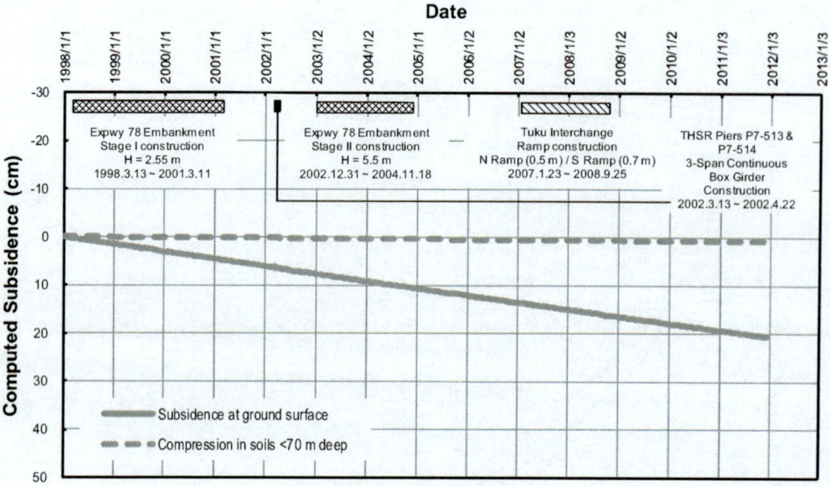

Fig. 12. Influence of soil creeping on compression of soils at the intersection of THSR and Expwy 78 between 01/1998 and 12/2011.

4.3 Influence of Overpumping on Compression of Soils

Figure 13 shows the influences of previous overpumping on the ground subsidence as well as the compression of soils with depths <70 m. Since previous overpumping is the most significant factor on the ground subsidence, the computed compression in soils with depths >70 m is also substantial, and would be the greatest among other factors as indicated in Table 2. Previous overpumping also produces appreciable compression in soils with shallower depths (<70 m). Numerical simulations indicate previous overpumping would be a key factor on both the distortion of THSR vertical alignment and the bearing capacity of THSR piles.

4.4 Influence of Expwy 78 Loading on Compression of Soils

Figure 14 shows the influences of Expwy 78 loading on the ground subsidence as well as the compression of soils with depths <70 m. Although Expwy 78 loading is a minor factor on the overall subsidence, the computed compression in shallower soils (<70 m), however, has become the greatest among other factors, as indicated in Table 2. In viewing that Expwy 78 embankment, with a width (B) of approximately 50m, exerted a strip load on the ground, considerable stress increments and compressions would be generated to a depth of around two times the strip width (i.e., 2B or 100 m). Alternatively, the computed compression in soils in deeper ground (>70 m or 100 m) would therefore be insignificant as compared with the one in shallower depths.

As the construction of Expwy 78 embankment at the intersection has been divided into three separated stages, the computed subsidence due to embankment loading responds in relation to the construction stages. It is noted that the second stage of construction (2003–2004) appears to be the most influential on the subsidence of ground as compared with those in the other two construction stages. Given the fact that the

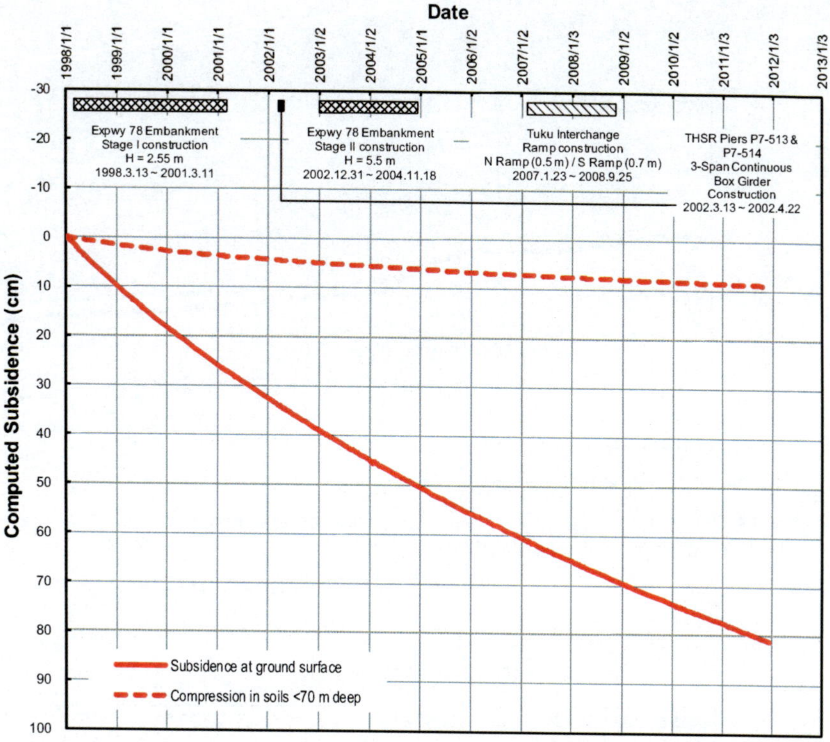

Fig. 13. Influence of previous overpumping on compression of soils at the intersection of THSR and Expwy 78 between 01/1998 and 12/2011.

placements of the first two construction stages of Expwy 78 embankment were nearly the same, the greater subsidence computed in the second stage would have appeared due to several possible reasons. The THSR pier loading that was in place approximately 9 months prior to the second construction stage would have been influential to enhance the subsidence. The construction period in the second stage (~2 years) was appreciably shorter than that of the first stage (~3 years), leading to a more intensive surcharge and hence greater subsidence of the ground. Groundwater fluctuations during the period of second construction stage appeared more severe than those of the first stage, and might have contributed to a more subsidence of the second stage construction.

4.5 Influence of THSR Loading on Compression of Soils

Figure 15 shows the influences of THSR loading on the ground subsidence as well as the compression of soils with depths <70 m. As mentioned previously, THSR structure exerts point load on the pile head and transmits through the shaft and tip of pile into the ground. In accordance, both the soils along pile shaft and below pile tip would be subjected to compression. Although THSR loading produces the least contribution to the total subsidence of the site, the compressions in soils with depths <70 m and >70

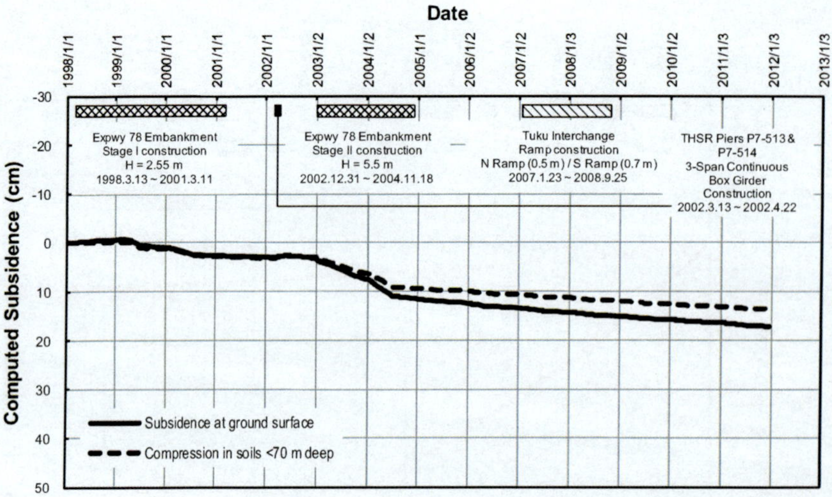

Fig. 14. Influence of Expwy 78 embankment loading on compression of soils at the intersection of THSR and Expwy 78 between 01/1998 and 12/2011.

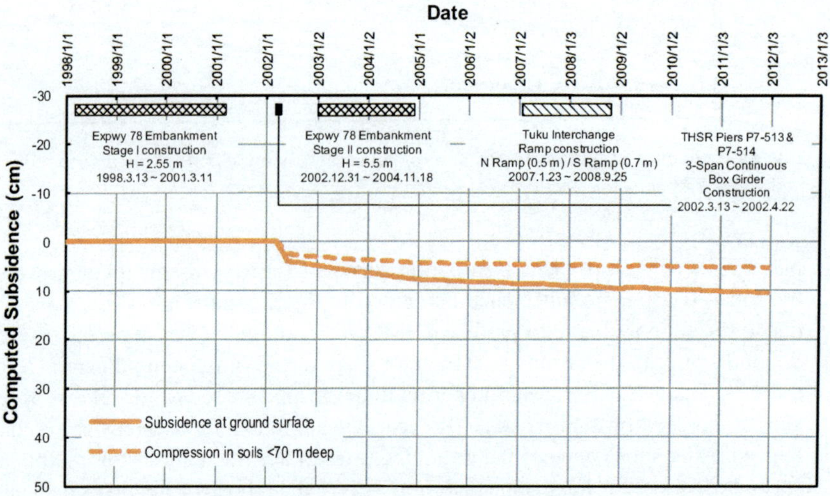

Fig. 15. Influence of THSR pier loading on compression of soils at the intersection of THSR and Expwy 78 between 01/1998 and 12/2011.

m would appear to be equivalent, as shown in the figure, due to characteristics of the loading.

4.6 Influence of GWT Fluctuation on Compression of Soils

Figure 16 shows the influences of groundwater fluctuations on the ground subsidence as well as the compression of soils with depths < 70 m. During numerical simulations, the groundwater fluctuations are assigned based on the observed data as shown in Fig. 8. However, limitations of the numerical software prohibit the use of complete sets of data with detailed measurements, but only allow for simplified versions of the groundwater measurements. Hence, the observed groundwater data is simplified and assigned with the average values in wet and dry seasons of the year. As seen in the figure, the ground subsides or rebounds in accord with the lowering or rising of groundwater levels, due to the changes in wet-dry seasons and the activities for agriculture farming. Groundwater fluctuations would have more influences on the compression of soils in deeper ground. For the shallower depths (<70 m), however, the compression in soils would be minor.

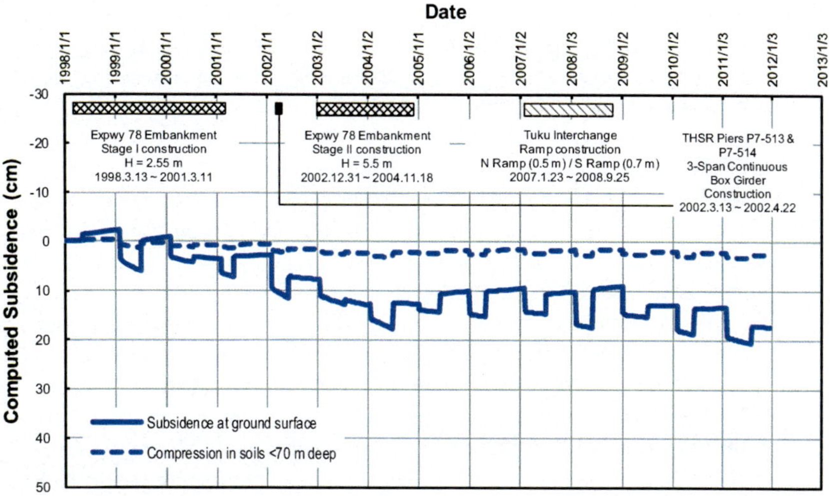

Fig. 16. Influence of groundwater level fluctuation on compression of soils at the intersection of THSR and Expwy 78 between 01/1998 and 12/2011.

5 Concluding Remarks

This paper presents results of numerical simulations on the subsidence behavior of ground at the intersection of THSR and Expwy 78, with emphases on the contributions of various factors on the subsidence as well as the influences of various factors on the compression of soils at shallower depths (i.e., < 70 m; the installation depth of THSR piles) and at deeper depths (i.e., > 70 m). The period for numerical simulations starts from 01/1998 and ends 12/2011, which covers full-ranges of constructions for THSR and Expwy 78 facilities at the intersection. In accordance, the results of simulations are able to reflect complete contributions on the subsidence by the loadings of these two transportation facilities. Some key findings of this study are summarized as follows:

- The subsidence of the study area is complicated by various influence factors, including: soil creeping, previous overpumping, groundwater fluctuation, and loadings of THSR viaducts/piers and Expwy78 embankment.
- For the 14-year period of concern (01/1998–12/2011), the numerical simulations reveal the total subsidence in the intersection area would be 147.0cm, or an average subsiding rate of 10.53 cm/year. The key contributing factor is found to be the previous overpumping (55.2%), followed by soil creeping (14.1%), groundwater fluctuation (11.8%), Expwy 78 loading (11.6%), and the least by THSR loading (7.3%), where the structural-related factors (i.e., THSR and Expwy 78 loadings) account for < 20% contribution on the total subsidence of the site.
- Numerical simulations indicate around 1/5 (or 21.4%) of total subsidence occurs as the compression in soils with depths < 70 m, and about 4/5 (or 78.6%) of total subsidence arises from the compression in soils with depths > 70 m. Accordingly, the compression rate in shallower soil deposits (<70 m) would be 2.25 cm/year, which would likely pose an unfavorable effect of negative skin friction and thus impair the bearing capacity of THSR piles installed within this depth range. Alternatively, the compression rate in deeper soil deposits (>70 m) would be 8.28 cm/year. The compression of deeper ground would settle THSR viaducts/piers and hence distort the vertical alignment of THSR route.
- For soil compressions in shallower depths (<70 m), the major influencing factor would be the Expwy 78 embankment loading, which accounts for 42.9% contribution of the compressions at this depth range. The Expwy 78 embankment exerts a static strip-load on the ground, and would be more pronounced on the stress increments and compressions as well in the shallower soil deposits.
- For soil compression in deeper ground (>70 m), the major influencing factor would be the previous overpumping, which accounts for 62.6% contribution of the compressions at this depth range. Previous overpumping that occurred approximately in the period of 1995–1997 significantly dropped the groundwater (or piezometer) levels, as evidenced by the monitoring data shown in Fig. 8, producing considerable and sustained straining on the ground mass. The influence of previous overpumping would be lasting for a long time until the piezometric levels of the onsite aquifers could be recovered and back to their original elevations.
- Numerical simulations reveal soil creeping would play an appreciable role on the subsidence of the study area (i.e., 14.1% of the total subsidence, or an average subsiding rate of 1.48 cm/year). Without groundwater infiltrations/evaporations or human activities like groundwater pumping and structural loading, the ground would still be sinking, as a result of secondary compressions (or creeping) of the enormously thick soil deposits of the CRAFD. Soil creeping occurs after the completion of primary consolidation, in a form of minor reorientations and slight adjustments of soil "grains" into a more stabilized arrangement. The primary consolidation due to natural deposition process of the soil deposits in CRAFD is believed to have completed for a long time, and the minor subsidence of ground at present, without the influences of groundwater infiltrations/evaporations and human activities, is due mainly to the creeping of the ground.
- As indicated previously, THSR facility would be suffered by the subsidence of the study area, causing negative skin frictions and reducing bearing capacity of THSR

piles due to the compression of soils in shallower depths (<70 m), as well as settling and distorting vertical alignment of THSR route as a result of the compression of soils in deeper ground (>70 m). The shallower depth compression is contributed mainly by the Expwy 78 embankment loading, while the deeper ground compression is contributed mainly by the previous overpumping. To ease the threat to THSR facility, this study suggests the identified key contributing factors, Expwy 78 embankment weight and previous overpumping, should be carefully dealt with. One potential mitigation measure would probably include a removal of earthen fill of the Expwy 78 embankment and a replacement with a light-weight structure, such as elevated road with box girders and pile foundation, EPS fill, etc. Other mitigation measure may also include a restriction of groundwater pumping and an enhancement of groundwater recharging, etc.

Acknowledgments. The authors would like to thank the funding provided by Resources Engineering Inc. Taiwan (NYUST 102–272) and Ministry of Science and Technology (previously, National Science Council), Taiwan (NSC102-2815-C-224-020-E, MOST105-2815-C-224-003-E). Some background information and monitoring data provided by Central Geological Survey, Water Resources Agency, and Directorial General of Highway, Taiwan, are highly appreciated.

References

Central Geological Survey (CGS), MOEA, Taiwan (1999). Summary report of groundwater monitoring network plan in Taiwan - phase I (1992–1998) (1999). (in Chinese)

Chang, M., Upomo, T.C., Huang, R.C., Wu, P.K., Su, P.C.: Evaluation on the behavior and contributing factors of subsidence in the intersection area of Expwy 78 and Taiwan High Speed Rail based on in-situ monitoring and numerical simulations. Bull. Eng. Geol. Env. **79**(6), 2971–2995 (2020)

Chappell, J., Shackleton, N.J.: Oxygen isotopes and sea level. Nature **324**, 137–140 (1986)

Chen, C.T., Hu, J.C., Lu, C.Y., Lee, J.C., Chan, Y.C.: Thirty-year land elevation changes from subsidence to uplift following the termination of groundwater pumping and its geological implication in the Metropolitan Taipei Basin, Northern Taiwan. Eng. Geol. **95**, 30–47 (2007)

Chen, C.H., Wang, C.H., Hsu, Y.J., Yu, S.B., Kuo, L.C.: Correlation between groundwater level and altitude variations in land subsidence area of the Choshuichi alluvial fan. Taiwan. Eng. Geol. **115**, 122–131 (2010)

Chia, Y.P., Lu, S.D., Wang, Y.S.: Hydrogeological framework of the south wing of Choshui River Fan. Proceedings of Conference on Groundwater and Hydrogeology of Choshui River Alluvial Fan, Taiwan, pp.113–125 (1996). (in Chinese with English Abstract)

Chiang, T.R., Lai, D.C., Huang, C.C., Lai, T.H., Chen, L.J. (1996). Conceptual model of hydrogeology and groundwater system of Choshui River Alluvial Fan. In: Proceedings of Conference on Groundwater and Hydrogeology of Choshui Alluvial Fan, pp.127–143 (1996). (in Chinese with English Abstract).

Ho, J.Y., Wang, G.C.: Political economics on groundwater extraction for the crisis of land subsidence in Yunlin Area. In: Proceedings of Annual Meeting on Cultural Studies, p. 23 (2012). (in Chinese)

Hung, W.C., Hwang, C., Chang, C.P., Yen, J.Y., Liu, C.H., Yang, W.: Monitoring severe aquifer-system compaction and land subsidence in Taiwan using multiple sensors: Yunlin, the southern Choshui River alluvial fan. Env. Earth Sci. **59**, 1535–1548 (2010)

Li, J.G.: Problems and mitigations of land subsidence - for areas in Yunlin and Changhua counties and along the THSR route. PPT of Keynote Speech at Geotech2013 Conference, September 11–13, Jenfushan Prince Hotel, Yunlin, Taiwan (2013). (in Chinese)

Sinotech: Calculation report of foundation, design unit: DU4.09, TK.224+259.842 ~ TK.224+404.842, Taiwan High Speed Rail. C270/DU4.09/DD, Sinotech Engineering Consultants, Ltd. (2003)

Water Resources Agency (WRA), MOEA, Taiwan (2011). Multi-sensors applied to monitor subsidence and investigate mechanism in Taipei, Changhua and Yunlin area in 2011 (in Chinese with English abstract), ISBN: 978-986-03-0205-9 (2011).

Water Resources Agency (WRA), MOEA, Taiwan (2019). Land subsidence map in Taiwan area in 2019 (2019). (in Chinese). https://www.wra.gov.tw/cp.aspx?n=3679.

Application of Recycled Rubber Mats for Improved Performance of Ballasted Tracks

Trung Ngo[1](✉), Buddhima Indraratna[2], and Cholachat Rujikiatkamjorn[1]

[1] Transport Research Center, School of Civil and Environmental Engineering, University of Technology Sydney, 15 Broadway, Ultimo, NSW 2007, Australia
Trung.Ngo@uts.edu.au

[2] Civil Engineering and Director, Transport Research Centre, School of Civil and Environmental Engineering, University of Technology Sydney (UTS), 15 Broadway, Ultimo, NSW 2007, Australia

Abstract. This paper summaries the results of a study on the use of recycled rubber mats for improved performance of ballasted tracks. One solution used to minimise ballast degradation (breakage) is to use an innovative recycled rubber mat, known as rubber-energy-absorbing drainage mat (READ), manufactured from end-of-life tires to provide a cost-effective solution to conventional tracks. When placed underneath the ballast, the energy-absorbing nature of the rubber mats decreases the load that is transferred to the ballast, so the ballast experiences less deformation and breakage. In this study, a series of large-scale triaxial tests are conducted to investigate the performance of the READS in the attenuation of cyclic and dynamic loads and subsequent reducing of ballast degradation. Numerical modelling using the Discrete Element Method (DEM) is conducted to investigate the improved performance of ballast in a micromechanical perspective. Evolutions of contact forces and contour stress distributions during cyclic tests are investigated through coupled DEM-FEM model.

Keywords: Recycled rubber mat · Transport geotechnics · Ballast · Breakage · Discrete Element Method (DEM)

1 Introduction

Australian rail sector must continuously upgrade its track network and adopt innovative technology to reduce the cost as well as increased passenger comfort. Ballast layer plays a crucial role in transmitting the train loading to underlying track-formation layers at a reduced stress level [1]. The ballast grains experience significant degradation and breakage when subject to heavy hauls freight trains and moving. Due to ballast breakage, Australian rail industry spends a significant budget in terms of track maintenance, together with improving the ground before track construction where soft and saturated subgrade soils induce considerable difficulties in design and construction [2].

The degradation of ballast aggregates (breakage) significantly influences the shear strength of ballast. Sun et al. [3] conducted large-scale tests on ballast and found that

© The Author(s), under exclusive license to Springer Nature Switzerland AG 2021
S. Shu et al. (Eds.): GeoChina 2021, SUCI, pp. 37–47, 2021.
https://doi.org/10.1007/978-3-030-80142-7_3

confining pressure (i.e., lateral confinement) and frequency have a significant influence on the breakage of ballast under cyclic loading. Leng *et al.* [4] conducted cyclic tests on coarse-granular aggregates and the test results showed that there were two separate deformation stages of a rapid increase in deformation initially followed by a steady-state of plastic strain after a given loading cycle. These findings were in agreement with previous literature [5–8], among others. Previous studies found that particle breakage could happen even at small-stress level [9, 10]. In addition, previous studies indicated that increased lateral confinement for ballast resulted in decreased peak friction angles [11].

Research on the application of elastic inclusions (geosynthetics, geocell) for improved performance of tracks and subsequently decreased ballast breakage has been well documented [12–18], among others. Field trials carried out on fully instrumented tracks at Singleton, NSW Australia proved that rubber mat could reduce ballast deformation apart from reduced breakage [19, 20].

There is a lack of study on the use of recycled rubber mats for reducing the breakage of ballast grains from a micro-mechanical perspective. This paper reports the main findings of the study on the use of large-scale triaxial tests and discrete element modelling (DEM) on the breakage responses of railway ballast.

2 Large-Scale Triaxial Cyclic Tests

A large-scale triaxial testing apparatus designed and built at the University of Wollongong can accommodate a testing sample of 300 mm diameter and 600 mm high, as detailed in Fig. 1a. Materials tested in the laboratory including ballast, capping materials and recycled rubber mat as shown in Figs. 1b–c. The ballast was cleaned and then passed through a set of standard sieves (53–9.5 mm) followed the Australian Standards (AS2758.7–2015) [21]. Upon drying, the ballast grains were coloured to help in quantifying the amount of breakage after each test. A mixture of sandy soil was adopted for a subgrade layer while a mixture of sand and crushed-rock was adopted for capping.

2.1 Sample Preparation and Testing Procedure

A tested sample was filled by a 200 mm thick subgrade at the bottom and compacted to a density of $\rho = 18.5$ kN/m^3. A 100 mm-thick capping was then compacted to $\rho = 20.5$ kN/m^3. A layer of rubber mats was then put on top of the capping. A ballast layer (300 mm thick) was placed and compacted to a $\rho = 15.5$ kN/m^3. After preparing the specimen, it was filled with water and left overnight to saturate the specimen. Cyclic tests were started under $q_{cyclic\text{-}max} = 230$kPa, $q_{cyclic\text{-}min} = 30$kPa, frequencies: $f = 10$–40Hz and a constant confining pressure of $\sigma_3 = 20$kPa. All tests were sheared up to $N = 500{,}000$ cycles. During these tests, the permanent vertical displacement, ε_a and volumetric strain, ε_v were measured. Ballast aggregates were recovered after each test, then sieved, and the changes in size distributions were recorded to investigate the amount of breakage.

Measured test data showed that the recycled rubber mats decreased deformation and breakage of ballast. This was because the mat could absorb cyclic and impact loads (i.e. energy) that led to less energy was transferred to ballast aggregates and thereby decreased breakage. The inclusion of rubber mat reduced the breakage up to 30.8% and 35.3% when subjected to frequencies of $f = 10$, 20 Hz, respectively. Detailed of test results were reported earlier by Indraratna *et al.* [22] and some of the test data were re-used in this study for the calibration of the DEM model.

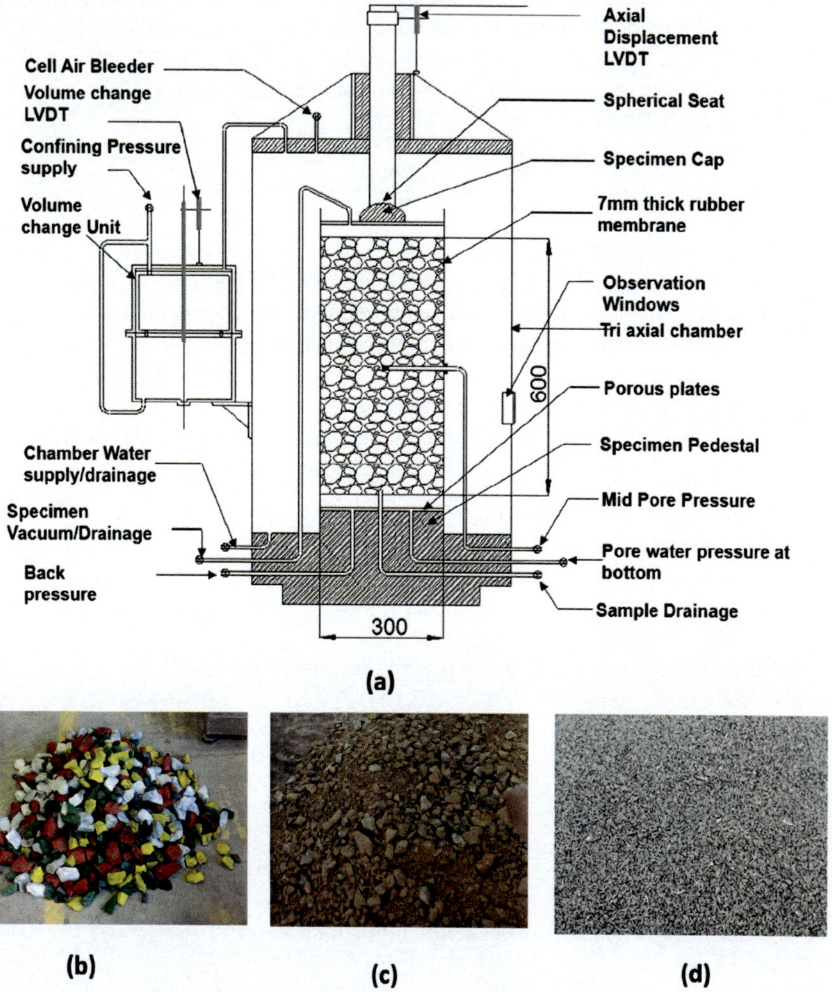

Fig. 1. (a) a schematic section of triaxial apparatus; (b) mixed ballast in different colours following the gradation; (c) capping aggregates; and (d) recycled rubber mat (READS)

3 Discrete Element Modeling

A discrete element method (DEM) has been widely used to study micro-mechanical responses of granular aggregates [23–28], among others. It is well documented that due to excessive computational time required to carry out cyclic tests in DEM, most of the previous DEM simulations were initiated to only a few thousand loading cycles. Given the excessively computational cost of simulating an actual track embankment in DEM having different layers of track substructure (i.e., ballast, capping, subgrade), thereby a coupled discrete-continuum modelling approach that utilises advantages of both numerical schemes with acceptable computational effort and accuracy is introduced in this study.

3.1 Coupled DEM-FEM

Ballast with varied geometries and sizes were digitalised in DEM by bonding many cylindrical together via parallel bond (Fig. 2a). The breaking (disconnection) of those bonds was approximately considered to represent particle breakage. Micro-mechanical parameters adopted for the DEM analysis were selected by calibration with laboratory test data as presented in Table 1. Figure 2b shows a schematic diagram of a coupled DEM-FEM model for the large-scale triaxial test, where the subgrade, capping and rubber mat was simulated by the continuum method. The schematic diagram of exchanging of forces (F_n, F_s) and displacement $\left(\dot{X}_i^{[E]}\right)$ between a discrete grain and continuum element at its interface is presented in Figs. 2c-d.

Contact forces at their interface can be calculated as:

$$F_i^{[c]} = F_n^{[c]} + F_s^{[c]} \quad (1)$$

where, normal force $\left(F_n^{[c]}\right)$ and shear force increment $\left(\Delta F_s^{[C]}\right)$ can be computed by:

$$F_n^{[C]} = K^n U^n n_i; \quad \text{and} \quad \Delta F_s^{[C]} = -K^s\left(\Delta X_i^{[C]} - \Delta X_i^{[C]} n_i\right) \quad (2)$$

Forces and moments on discrete particles can be determined by:

$$F_i^{[P]} \leftarrow F_i^{[P]} - F_i^{[C]}; \quad \text{and} \quad M_i^{[P]} \leftarrow M_i^{[P]} - e_{ijk}\left(X_j^{[C]} - X_j^{[P]}\right)F_k^{[C]} \quad (3)$$

Relative velocity (displacement) at the interface (V_i) is given by:

$$V_i = \dot{X}_{i,E}^{[C]} - \dot{X}_{i,E}^{[P]} = \dot{X}_{i,E}^{C} - \left[\dot{X}_i^{[P]} + e_{ijk}\omega_j^{[P]}\left(X_k^{[C]} - X_k^{[P]}\right)\right] \quad (4)$$

where, $\dot{X}_{i,E}^{[C]}$and $\dot{X}_{i,E}^{[P]}$ are the velocities of elements and discrete grains. $\dot{X}_i^{[P]}$and $\omega_j^{[P]}$are the translation and rotation of a grain, and e_{ijk} is the permutation symbol.

The velocity of continuum elements at the interface is then calculated as:

$$\dot{X}_{i,E}^{[C]} = \sum N_j \dot{X}_{i,E}^{j} \quad (5)$$

where, $N_j$is a shape function, determined by: $N_j = (1 + \xi_o)(1 + \eta_o)/4, j = 1, 2, 3, 4;$ *and* $\xi_o = \xi_i\xi, \eta_o = \eta_i\eta$; ξ_i and η_i are coordinates of nodes. Shear and normal forces are distributed to the nodal force, $F_i^{[E,j]}$ via shape function N_j:

$$F_i^{[E,j]} = F_i^{[E]} + F_i^{[C]}N_j \tag{6}$$

During load application, contact forces DEM domain (Zone 1) is transferred to continuum domain (Zone 2); and the displacements of continuum elements, are transferred back to the DEM – Zone 1 as boundary conditions. More detail on the model calibration, application of loading and boundary conditions, and laboratory test results can be found in [22].

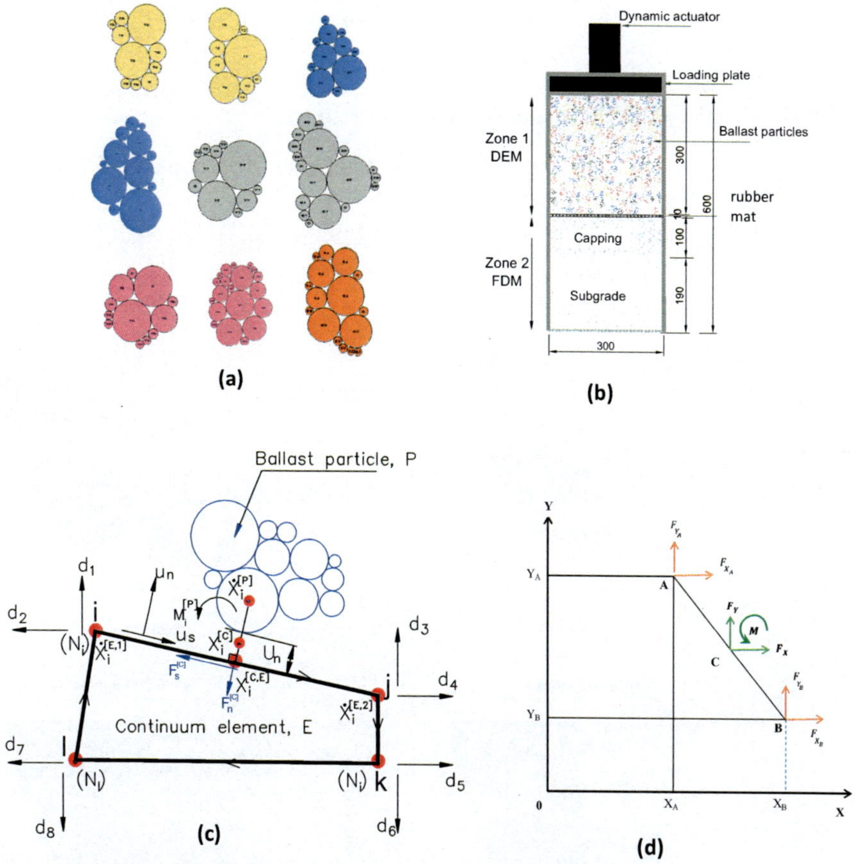

Fig. 2. (a) Simulated ballast grains; (b) coupled DEM-FEM model; and (c) interaction of a grain and an element; and (d) force and moment exchanges (*modified after* [24])

Table 1. Micro-mechanical parameters adopted for DEM analysis

Parameter	Value
Particle density (kg/m^3)	2500
Friction coefficient	0.82
Contact normal stiffness (N/m)	2.52×10^8
Contact shear stiffness (N/m)	1.26×10^8
Contact normal stiffness of wall (N/m)	1.25×10^8
Shear stiffness of wall (N/m)	0.625×10^8
Parallel bond stiffness (N/m)	6.84×10^{10}
Parallel bond shear strength (N/m^2)	486×10^4
Parallel bond radius multiplier	0.50

3.2 Predicted Stress-Strain Behaviour of Ballast

Figure 3 presents predicted cyclic stress-strain from the coupled DEM-FEM model at varied load cycles N subjected to a frequency, $f = 20$ Hz. The predicted strains remarkably increase up to about 5.5% at the first N = 1000 load cycles, followed by

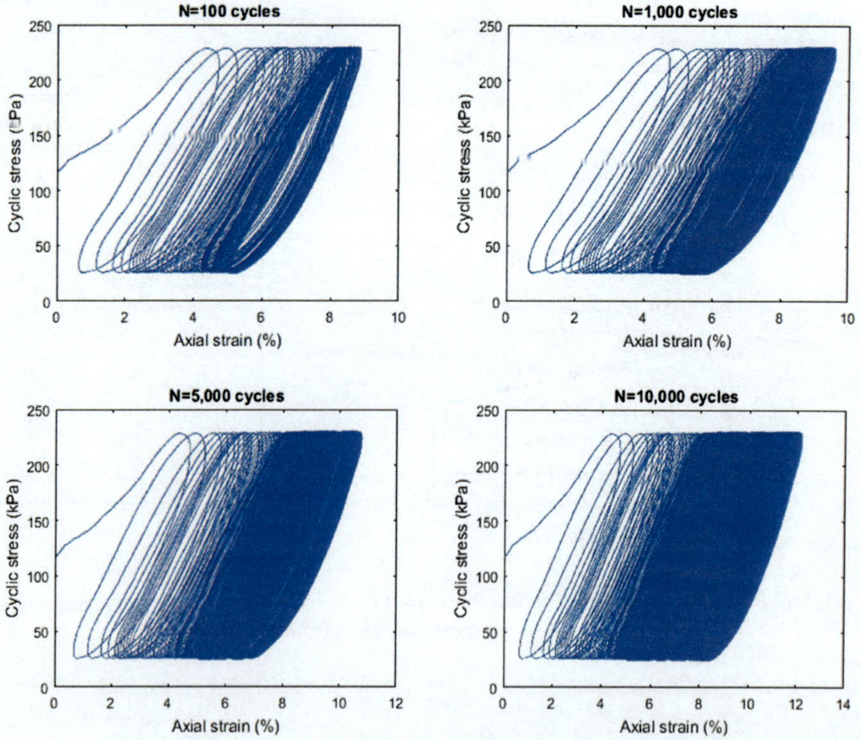

Fig. 3. Applied cyclic stress versus axial strain ε_a at varying load cycle, N

slightly increased strains up to N = 5,000 cycles. After this cycle, the ballast remains relatively unchanged to the end of the simulation (N = 10,000 cycles). This indicated that ballast aggregates exhibited remarkable compression and re-arrangement during the initial stage of loading, but after achieving a threshold compression, any subsequently applied loads would resist further deformation and instead, accelerate ballast breakage.

3.3 Simulated Ballast Breakage

The accumulation of bond breaking with increased N subjected to 4 frequencies of $f = 10–40$ Hz is presented in Fig. 4. It is noted that the breaking of bonds (B_r) within a simulated particle was considered to represent particle breakage. It is worth mentioning that B_r is different from the actual ballast breakage measure in the laboratory, the B_r could supply an indication of the extent of breakage occurs. From DEM-FEM simulation, the B_r increases with an increase in load frequency, f and this trend are similar to the ballast breakage measured in the laboratory data. Subjected to a given f, the number of broken bonds increases significantly within the first $N = 5{,}000$ cycles; and then keeps

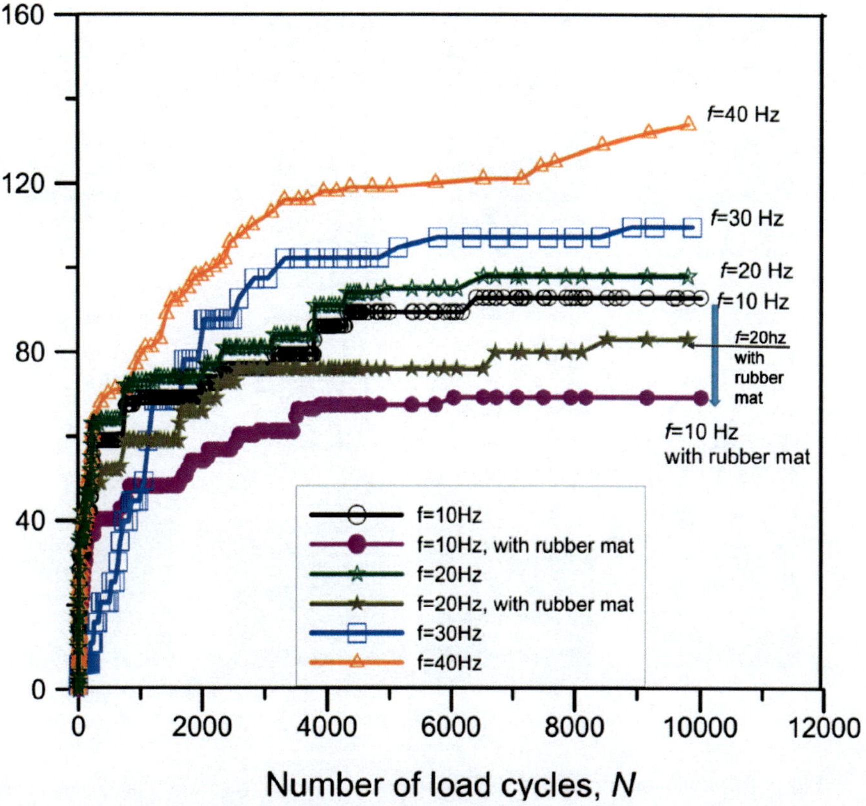

Fig. 4. Predicted contact bond breaking at different frequencies, $f = 10–40$ Hz with and without recycled rubber mat *(adopted from Indraratna et al. 2020 – with permission from ASCE)*.

almost unchanged at subsequent loading cycles. The inclusion of a recycled rubber mat underneath the ballast layer leads to a decreased B_r; and this could be attributed to the energy-absorbing characteristics of the rubber mat.

3.4 Contact Force Distributions

Figure 5 presents the distributions of contact forces in the ballast layer and vertical stresses predicted for capping and subgrade layers. Vertical stress contours in the capping and subgrade layers at different load cycles N are also presented where the contour was captured under $f = 20$ Hz. It is seen that the applied cyclic loads are distributed to particles through contact force chains where each contact force is represented at the contact point having a thickness proportional to the force magnitude. Upon cyclic loading, large contact forces appeared and concentrated beneath a top-loading plate. It

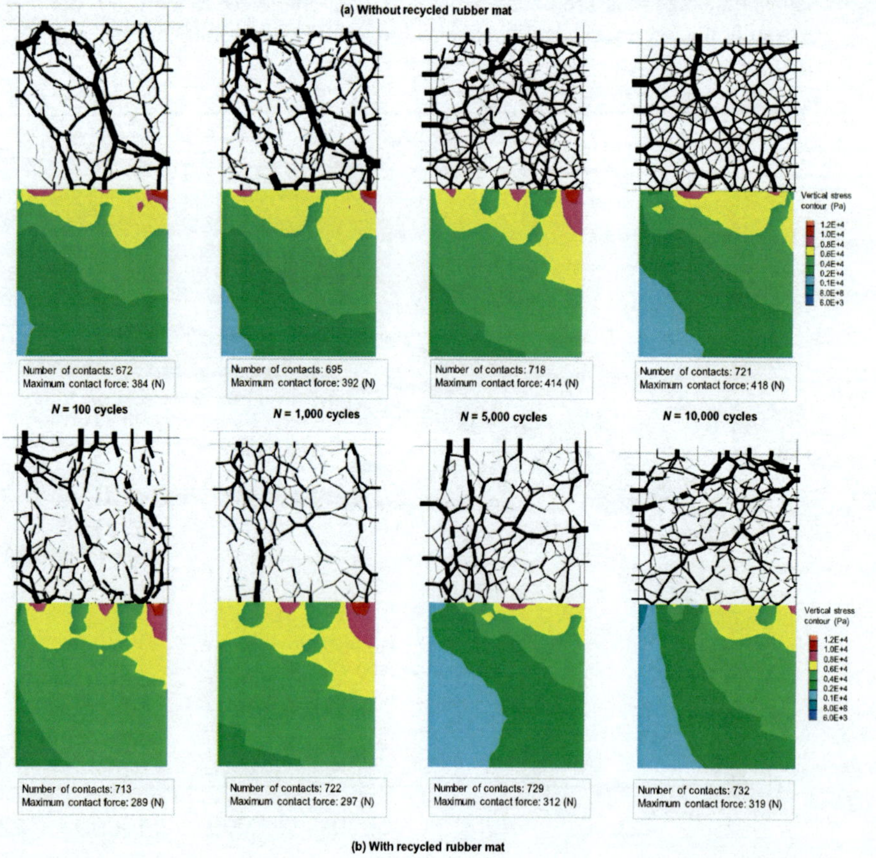

Fig. 5. Distributions of contact forces in ballast layer and vertical stresses predicted for capping and subgrade: (a) without rubber mat; and (b) with a rubber mat *(adopted from Indraratna et al. 2020)*.

is clearly seen that the inclusion of rubber mat could increase the contact areas among ballast particles and underneath layers, widen the distribution of vertical stresses; and this in turn decrease magnitude of contact forces the interfaces. Predicted vertical stress at the capping is large at the interfaces with the ballast grains, whereby it is predicted to significantly decrease with depth. An increased applied load cycle N, leads to an increased number of contact force and increased contact force magnitude. This could be associated with the densification and re-arrangement of particles.

4 Conclusions

The following findings can be established:

- Measured laboratory test data proved that the axial deformation, ε_a increased with an increased N and the higher frequency applied the higher ε_a was measured. Ballast grain showed a large axial strain, ε_a within the first $N = 1000$ cycles, followed by a slight increase in the ε_a at a decreased rate.
- The inclusion of a recycled rubber mat resulted in decreased axial strain and breakage of ballast. This was because the mat absorbed cyclic and impact loads and thereby reduced particle breakage of up to 30.8% and 35.3% when subjected to $f = 10, 20$Hz.
- A coupled DEM-FEM model was introduced to investigate the applied cyclic load-deformation responses of ballast, where discrete grains were modelled by the DEM and the continuum media (REAM, capping and subgrade) were modelled by a continuum media. From the coupled discrete-continuum model (coupled DEM-FEM), the predicted amount of broken bonds (i.e., mimicking particle breakage) increased with increased frequency ($f = 10$–40 Hz) and it decreased with the inclusion of a rubber mat.
- The force distributions were studied, and they showed that the forces distributed non-uniformly across the ballast specimens and the inclusion of a rubber mat decreased the magnitude of contact forces which also implied for a reduced particle breakage.

Acknowledgements. The authors acknowledge funding provided by the Australian Government through the ITTC-Rail, (IC170100006). The authors are grateful to the support from organizations including RM CRC and Australasian Centre for Rail Innovation (ACRI) through Project R2.5.1.

References

1. Raymond, G.P., Davies, J.R.: Triaxial tests on dolomite railroad ballast. J. Geotech. Eng. Div. Proc. Am. Soc. Civil Eng. **104**(GT6), 737–751 (1978)
2. Ravitharan, S.S.S., Martin, M.A., Minness, C.: Understanding load tranfer characteristics withn track sub-structure: queensland experience. In: Conference on Railway Engineering Proceedings: Engineering Innovation for Competative Edge, Queensland (1998)
3. Sun, Q., Indraratna, B., Ngo, N.T.: Effect of increase in load and frequency on the resilience of railway ballast. Géotechnique **69**(9), 833–840 (2019)

4. Leng, W., Xiao, Y., Nie, R., Zhou, W., Liu, W.: Investigating strength and deformation characteristics of heavy-haul railway embankment materials using large-scale undrained cyclic triaxial tests. Int. J. Geomech. **17**(9), 04017074 (2017)
5. Suiker, A.S.J., Borst, R.: A numerical model for the cyclic deterioration of railway tracks. Int. J. Numer. Meth. Eng. **57**, 441–470 (2003)
6. Indraratna, B., Ngo, N.T., Rujikiatkamjorn, C.: Deformation of coal fouled ballast stabilized with geogrid under cyclic load. J. Geotech. Geoenviron. Eng. **139**(8), 1275–1289 (2013)
7. Jayasuriya, C., Indraratna, B., Ngo, T.: Experimental study to examine the role of under sleeper pads for improved performance of ballast under cyclic loading. Transp. Geotech. **19**, 61–73 (2019)
8. Xiao, Y., et al.: Effects of load duration and stress level on deformation and particle breakage of carbonate sands. Int. J. Geomech. **20**(7), 06020014 (2020)
9. Xiao, Y., Liu, H., Chen, Y., Jiang, J.: Strength and deformation of rockfill material based on large-scale triaxial compression tests. I: influences of density and pressure. J. Geotech. Geoenviron. Eng. **140**(12), 04014070 (2014)
10. Ueng, T.-S., Chen, T.-J.: Energy aspects of particle breakage in drained shear of sands. Géotechnique **50**(1), 65–72 (2000)
11. Indraratna, B., Ngo, T.: Ballast Railroad Design: Smart-Uow Approach. CRC Press, Boca Raton (2018)
12. Zhai, W.M., Wang, K.Y., Lin, L.H.: Modelling and experiment of railway ballast vibrations. J. Sound Vib. **270**, 673–683 (2004)
13. Powrie, W., Yang, L.A., Clayton, C.R.I.: Stress changes in the ground below ballasted railway track during train passage. Proc. Inst. Mech. Eng. Part F: J. Rail Rapid Transit 247–261 (2007)
14. McDowell, G.R., Lim, W.L., Collop, A.C., Armitage, R., Thom, N.H.: Comparison of ballast index tests for railway trackbeds. Geotech. Eng. **157**(3), 151–161 (2008)
15. Tutumluer, E., Huang, H., Bian, X.: Geogrid-aggregate interlock mechanism investigated through aggregate imaging-based discrete element modeling approach. Int. J. Geomech. **12**(4), 391–398 (2012)
16. Ngo, N.T., Indraratna, B., Rujikiatkamjorn, C.: DEM simulation of the behaviour of geogrid stabilised ballast fouled with coal. Comput. Geotech. **55**, 224–231 (2014)
17. Leng, J., Gabr, M.A.: Deformation-resistance model for geogrid-reinforced. Transp. Res. Rec. **1975**, 146–154 (2006)
18. Navaratnarajah, S.K., Indraratna, B., Ngo, N.T.: Influence of under sleeper pads on ballast behavior under cyclic loading: experimental and numerical studies. J. Geotech. Geoenviron. Eng. **144**(9), 04018068 (2018)
19. Priest, J.A., Powrie, W., Yang, L., Grabe, P.J., Clayton, C.R.I.: Measurements of transient ground movements below a ballasted railway line. Geotechnique **60**(9), 667–677 (2010)
20. Indraratna, B., Nimbalkar, S.S., Ngo, N.T., Neville, T.: Performance improvement of rail track substructure using artificial inclusions – experimental and numerical studies. Transp. Geotech. (2016). https://doi.org/10.1016/j.trgeo.2016.04.001
21. 2758.7, A Aggregates and rock for engineering purposes, Part 7. Railway Ballast. Standard Australia, 2015. NSW, Australia (2015)
22. Indraratna, B., Ngo, T., Rujikiatkamjorn, C.: Performance of ballast influenced by deformation and degradation: laboratory testing and numerical modeling. Int. J. Geomech. **20**(1), 04019138 (2020)
23. McDowell, G.R., Li, H.: Discrete element modelling of scaled railway ballast under triaxial conditions. Granular Matter **18**(3), 1 (2016). https://doi.org/10.1007/s10035-016-0663-8
24. O'Sullivan, C., Cui, L., O'Neill, C.: Discrete element analysis of the response of granular materials during cyclic loading. Soils Found. **48**(4), 511–530 (2008)
25. Aursudkij, B., McDowell, G.R., Collop, A.C.: Cyclic loading of railway ballast under triaxial conditions and in a railway test facility. Granular Matter **11**, 391–401 (2009)

26. Huang, H., Tutumluer, E., Hashash, Y.M.A., Ghaboussi, J.: Discrete element modeling of coal dust fouled railroad ballast behavior. In: The 89th Annual Meeting DVD of the Transportation Research Board, Washington, D.C. (2010)
27. Jiang, H., Bian, X., Cheng, C., Chen, Y., Chen, R.: Simulating train moving loads in physical model testing of railway infrastructure and its numerical calibration. Acta Geotech. **11**(2), 231–242 (2014). https://doi.org/10.1007/s11440-014-0327-y
28. Ngo, N.T., Indraratna, B., Rujikiatkamjorn, C.: A study of the geogrid-subballast interface via experimental evaluation and discrete element modelling. Granular Matter **19**(3), 1–16 (2017)

Appraisal of Railway Ballast Degradation Through Los Angeles Abrasion, Cyclic Loading Tests, and Image Technics

Zhihong Nie[1], Mohammed Ashiru[1,2](✉), Xingchen Chen[1], and Said Hussein Mohamud[1]

[1] School of Civil Engineering, Central South University, Changsha 410075, China
mohammeda@abu.edu.ng
[2] Ahmadu Bello University, Zaria 810107, Nigeria

Abstract. Ballast degrades through wear and breakage due to the cumulative actions of traffic. This study evaluated ballast degradation using the Los Angeles abrasion (LAA) and cyclic loading tests, changes in the ballast aggregates shape indexes (form, angularity, surface texture and volume) were also evaluated using image analysis techniques. A typical freight train with an average speed of 120 km/h and 30 tonnes capacity was simulated for the cyclic loading test. Series of LAA and cyclic loading tests were conducted on separate samples of the same gradations, and the changes in the aggregates shape indexes were determined using the Aggregate image measuring system (AIMS) package and a 3D laser scanner before and after each of the tests. The image-based assessment revealed that ballast degradation is related to the particle edge breakage and surface wear. The overall experimental results showed that ballast degradation is more profound in the ballast early stage of service, and the degradation index has a good correlation with the changes in aggregate shape indexes. The results also revealed that the two tests gave different degradation indexes, hence a relationship between the ballast degradation index and the number of load cycles with the number of LAA drum turns is proposed to reflect the effect of the number of train's axle load cycles on ballast degradation.

1 Introduction

Ballast degradation through attrition and breakage during operations affects the structural performance of the railway track system. Conventionally, the ballast is designed to; "provide a solid bed for the sleepers, distribute stresses from the sleeper bearing area to a bearable limit at the surface of the subgrade soil, facilitate maintenance operations for the track riding quality, and provide rapid drainage" (Lu and McDowell 2010). Over time, ballast degrades and foul due to heavy haul trains cyclic loading and maintenance tamping. The edges of the ballast particles break and rearrange resulting in settlement of the track system, hence reducing the volume of voids (Han 2012; Indraratna *et al.* 2014). Attrition of ballast begins at the construction stage during tamping, produces pulverized ballast particles that continue after the railway is open to traffic, and subject

© The Author(s), under exclusive license to Springer Nature Switzerland AG 2021
S. Shu et al. (Eds.): GeoChina 2021, SUCI, pp. 48–58, 2021.
https://doi.org/10.1007/978-3-030-80142-7_4

to environmental conditions that further results in the ballast degradation and fouling of the ballast voids (Danesh et al. 2018; Guo et al. 2019 and Qian et al. 2017).

With the railway sector contributing immensely to the continuous rapid economic growth of China and the high demand for Chinese products, the Chinese government intensified effort to continue expanding its railway system to its full capacity (Cui et al. 2018). This, for instance, will improve the transportation of coal from a mining site to generation plants and manufactured products to national export points and other international destinations. To meet up with the demand, the annual tonnage increases and lead to more numbers of traffic load cycles. Hence, the ballast deteriorates progressively due to increasing load cycles (Indraratna et al. 2016). Studies on railway ballast degradation have been conducted using different approaches; Los Angeles abrasion test, Crushing test, and cyclic loading test as reported in the literature. McDowell et al. (2005) conducted a study in a Laboratory by simulating the effects of both train loading and ballast tamping on four granitic rocks. The results revealed that the amount of ballast degradation agrees with that of the Los Angeles Abrasion (LAA) test. Aursudkij (2007) conducted similar research on three different types of rock. He concluded that Los Angeles Abrasion (LAA) test results correlate well with ballast degradation under cyclic loadings. Röthlisberger et al. (2006) reported that Los Angeles Abrasion (LAA) test results correlate well with ballast degradation induced by railway traffic. Nålsund et al. (2013) conducted a comparative assessment of ballast degradation using a large-scale triaxial and full-scale rail track model tests with empirical laboratory test methods on Norwegian ballast materials. They reported that the LA abrasion test results correlated with the actual material breakdown. Qian et al. (2014), and (2017) characterized ballast aggregate degradation using LA abrasion tests and aggregate image analysis. The results revealed that with a decrease in particle angularity and surface texture, the percentage of fines increased. Hence a good correlation between the ballast fouling index and the aggregate morphological changes was observed. Sun et al. (2014) reported from their study that ballast degradation correlates with its volumetric shrinkage under cyclic loading conditions.

Although researchers have been working on quantifying ballast degradation using different approaches (empirical and mechanistic methods), studies relating the two approaches have been relatively scarce in the literature. Developing a relationship between the two approaches would enable railway engineering practitioners to ascertain the number of load cycles at which the rail track will be due for maintenance using the preliminary LAA test result data of the ballast material. In view of this, the current study aims at; (i) Evaluating ballast degradation using the LAA test method that is purely empirical, and using the cyclic loading test method that is mechanistic. (ii) Ascertaining changes in the ballast particles morphological properties under different loading conditions. (iii) Carrying out a comparative analysis of results from the two tests and propose a relationship between the number of train-axle-load cycles and ballast degradation.

2 Material and Methods

2.1 Material

The ballast material used for this study is a 100% crushed granite collected from Xiangxiang quarry about 114 km away from Changsha in Hunan province, China. The aggregates with different particle sizes were carefully selected, washed, oven-dried, and air-dried at room temperature. The specimens were prepared by mixing the aggregates according to the target proportion, as per the People's Republic of China Railway Industry-standard gradation (TB/T 2328.14-2008) requirements as shown in Fig. 1.

2.2 Methods

2.2.1 Los Angeles Abrasion (LAA) Test

To simulate the degradation of ballast particles that is closer to the real situation on the field, Fischer (Fischer 2018) suggested a Los Angeles abrasion test without steel balls. He argued that in actual circumstances ballast particles are not subject to adverse stresses in the field like the one induced by the steel balls. The Los Angeles abrasion (LAA) test was performed in accordance with ASTM Standard C535 (ASTM 2009) but without the steel balls. The drum rotates at an average speed of 33 revolutions per minute, and for each test, the drum rotated 250 times. Afterward, material passing each sieve was collected and recorded. Particles larger than 9.5 mm were collected for image analysis using the Aggregate Image Measurement System (AIMS) and 3D laser scanner.

2.2.2 Cyclic Loading Test

The specimens for the cyclic loading tests were prepared in accordance with the gradation requirement and compacted to a density of 1512 kg/m^3. A typical heavy haul freight train with an average speed of 120 km/h corresponding to a frequency of 16.5Hz obtained as 0.138V (Indraratna *et al.* 2014; Sun *et al.* 2014) and axle load of 30 tons which corresponds to 288 kPa (Esveld and Esveld 2001) was simulated. Series of cyclic loading tests were conducted at varying load cycles from 200000 – 1000000 cycles at an interval of 200000. Tests were conducted with maximum deviator stress of 288 kPa and the minimum stress was set to 45 kPa for all tests. The specimens were sieved after loading using standard sieves, material passing each sieve was collected and recorded. Particles larger than 9.5 mm were collected for image analysis using the Aggregate Image Measurement System (AIMS) and some selected ones were scanned using a 3D laser scanner. The ballast degradation index was estimated using Marsal's Breakage index (B_g) (Marsal 1967).

Marsal's breakage index (B_g): the breakage index proposed by Marsal (Marsal 1967) is determined as the summation of positive values of the difference in percentage by mass of total material retained on the same sieve size after the test (Koohmishi and Palassi 2018). The differences in percentage retained on each sieve size (ΔW_k) is determined as follows;

$$\Delta W_k = W_{ki} - W_{kf} \tag{1}$$

Where; W_{ki} and W_{kf} ; are the percentages by mass retained on sieve size k before and after the test respectively. Marsal's breakage index (B_g) is adopted for this study because of its simplicity and accuracy (Indraratna and Salim 2002).

2.2.3 Image Analysis

AIMS software was used to determine the essential morphological indices (are the angularity index (AI), the surface texture (ST) index, and the flat and elongated (F&E) index). These three key indices are as illustrated in Fig. 3. AIMS angularity index quantifies the total changes in the particle edge sharpness characteristics on a scale of 0–10000. It is categorized into four groups (low, moderate, high, and extremely angular), as illustrated in Fig. 4. The F&E index quantifies the index of the longest to the shortest dimensions of a particle in an assembly of flat, elongated, cubical particles. The ST index quantifies the roughness of a particle through an erosion-dilation image analysis technique on a scale of 0–1000. It is also categorized into four groups, as illustrated in Fig. 5. The image analysis was conducted to determine the changes in particle size and shape after each test before mixing all materials for the next tests (Fig. 2).

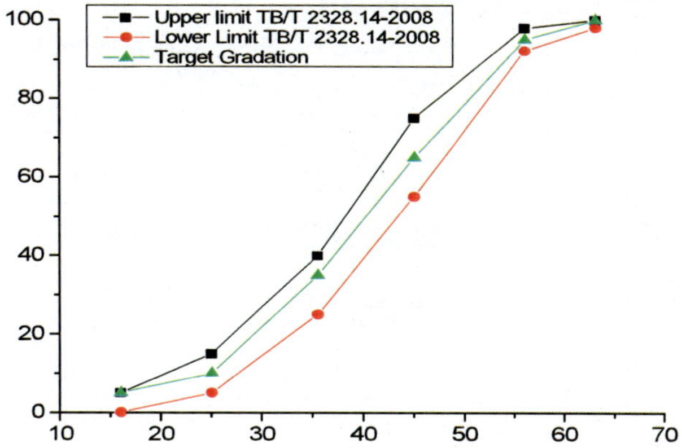

Fig. 1. Gradation of crushed granite ballast material (TB/T 2328.14–2008)

3 Results and Discussion

3.1 Degradation

The gradation curves obtained from the sieve analysis of the ballast after 2000 turns of the LA abrasion drum at an interval of 250 turns are as presented in Fig. 6. The figure also presents the corresponding values of the breakage index obtained after every 250 turns of the LAA test. It can be seen that with an increase in the number of turns of the LA abrasion test drum, the specimen's gradations shift gradually to the right, and the breakage index increases at a decreasing rate. Figure 7 presents the changes in gradation after the cyclic loading test with the corresponding breakage indices at an interval of 200000 load cycles. A similar trend to that of the LAA test was observed; an increase in the number of load cycles yields an increase in breakage index. It is clearly shown that with an increase in the number of LAA drum turns and load cycles, the breakage index (B_g) values increase. The increase in breakage index could be attributed to the breakage

Fig. 2. Second-generation of aggregate imaging measurement system (AIMS)

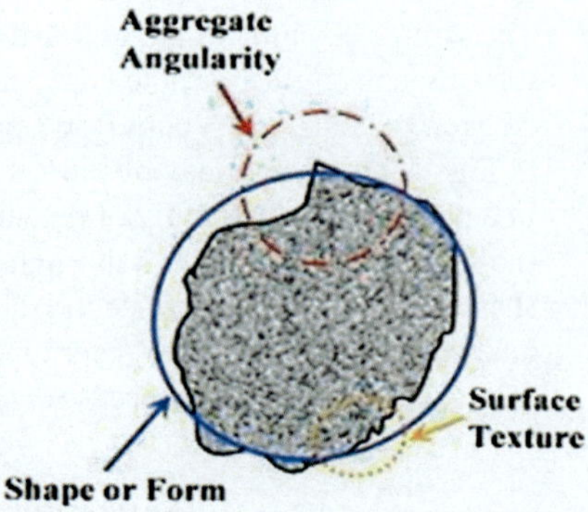

Fig. 3. Key morphological descriptors (Masad and Fletcher 2005)

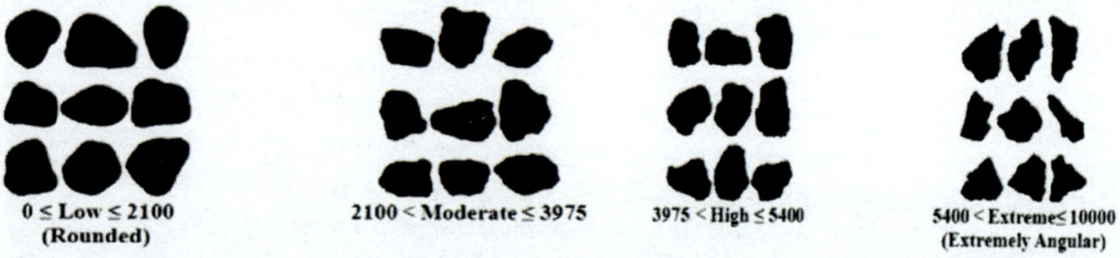

Fig. 4. Particles angularity

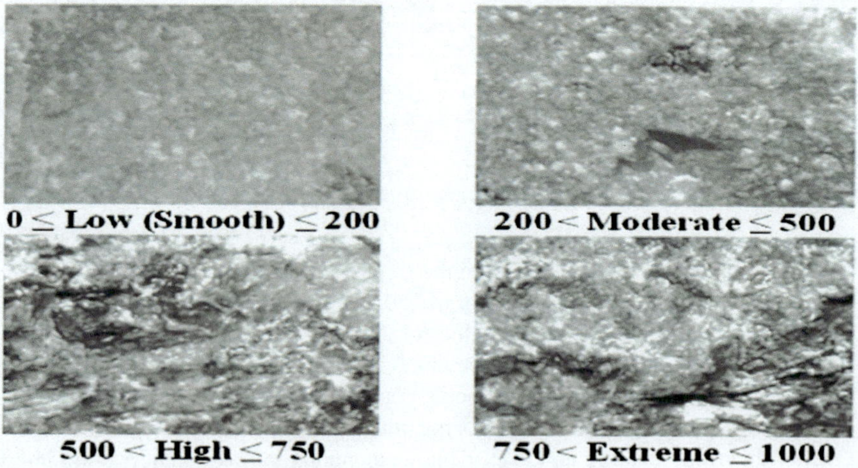

Fig. 5. Particles surface texture

and splitting of larger particles into smaller ones. Hence, the particles become rounded due to wear and abrasion of the particles' sharp edges with an increase in the number of load cycles/LAA drum turns. Therefore, ballast degradation in the field is attributed primarily to the breakage of sharp corners during service. These findings agree with the works of Indraratna et al. (2014); (2013)and Qian *et al.* (2017).

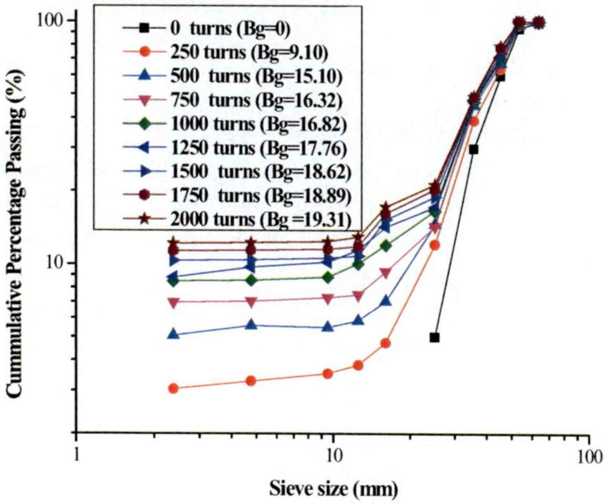

Fig. 6. Gradation changes of ballast specimen after LA abrasion test

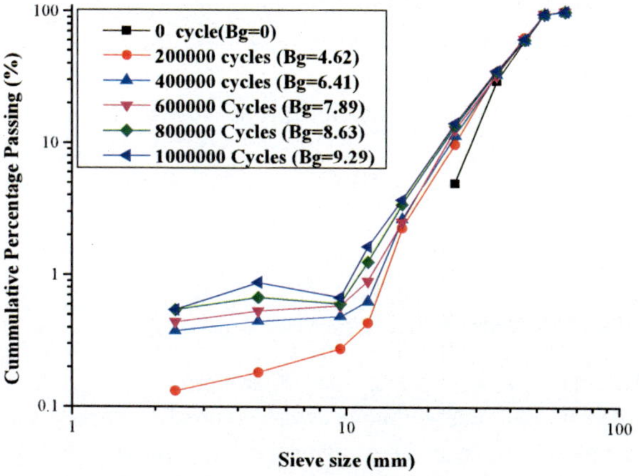

Fig. 7. Gradation changes of ballast specimen after cyclic loading test

Figures 8 and 9. Presents the ballast degradation trends in terms of breakage index (B_g) values against the number of LAA drum turns and the number of load cycles respectively. Figure 8 presents the relationship between the breakage index (B_g) and the number of turns of the LAA (N_{LAA}) test. The result indicates that with an increase in N_{LAA}, the breakage index also increases, which implies that the particles keep degrading through sharp corners breakage, attrition of asperities, and splitting of larger particles. Based on the data presented in Fig. 8. A best-fit equation with a coefficient of determination (R^2) of 99.56%, the data can be represented in Eq. (2 and 3).

$$B_g = 17.0233 + 1.2347N_{LAA} - 32611.2014N_{LAA}^{-1.5} \tag{2}$$

$$N_{LAA} = 0.9728e^{0.3921B_g} \tag{3}$$

Where; B_g is the breakage is index and N_{LAA} is the number of turns of the LAA test drum.

The relationship between breakage index (B_g) and the number of load cycles for the cyclic loading test is as illustrated in Fig. 9. The test result revealed that with an increase in the load cycles, the breakage index of the particles also increases. Hence, it indicated that the ballast continues to degrade with an increase in load cycles. Therefore, in order to ascertain ballast degradation in the field during service, a relationship between breakage index (B_g) and the number of load cycles based on the data presented in Fig. 9 is proposed as in (4 and 5). The proposed relationship could help railway practitioners to plan for railway ballast maintenance.

$$\mathrm{B_g} = -30.0943 + 1.8286 \times 10^{-13}\left(\mathrm{N_{cyc}^2}\right) + 2.8397\ln\left(\mathrm{N_{cyc}}\right) \tag{4}$$

$$N_{CYc} = 76381.52e^{0.2857B_g} \tag{5}$$

Where; B_g is the breakage index and $\mathrm{N_{cyc}}$ is the number of load cycles.

However, from the data presented in Figs. 8 and 9, 10. Illustrates a proposed relationship between the number of turns of the LAA test drum and the number of load cycles. This relationship can also be presented in (6). This relationship could aid practitioners in predicting the load cycles at which the railway ballast is due for maintenance.

$$\mathrm{N_{Cyc}} = 4.79 \times 10^8 - 1.76 \times 10^7 e^{(-4.58\times10^{-4}N_{LAA})} - 4.63 \times 10^8 \mathrm{e}^{(-1.04\times10^{-5}N_{LAA})} \tag{6}$$

3.2 Imaging Assessment

The image analysis was conducted before and after each test for both LAA and cyclic loading tests. The imaging techniques were employed to assess the changes in the particle morphological properties in terms of three independent indexes; form (flatness and elongation), angularity, and surface texture. The changes in morphological features of ballast aggregate were characterized during the degradation process using the AIMS package, and a 3D laser scanner was used to capture the 3D images in order to evaluate the changes in the particle volume and surface area. Table 1 illustrates the changes in the AI angularity index, ST Surface texture, flat and elongated index, Volume, and surface area of a single aggregate particle at intervals of the number of LA abrasion drum turns. It appears that with an increase in number of LA abrasion drum turns, the particles lose their asperities, sharp corners, and become rounded, and hence, all the image-based

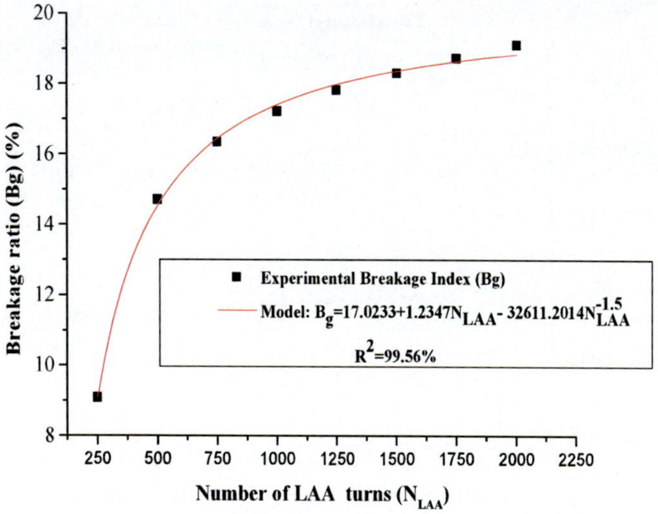

Fig. 8. Relation between the breakage index and the number of turns of the LAA test drum.

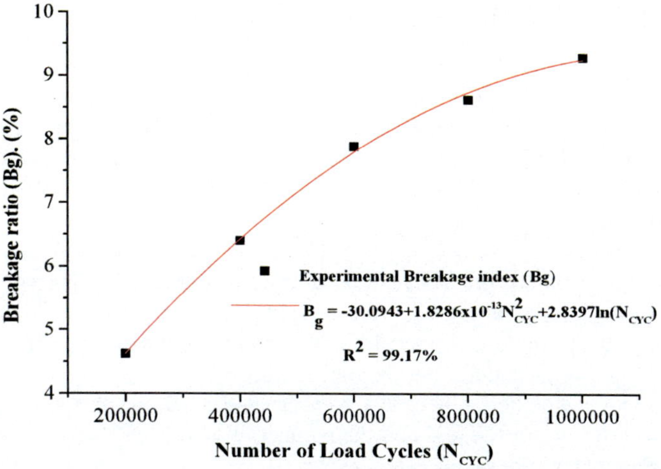

Fig. 9. Relation between the breakage index and the number of load cycles.

indexes decrease. This conforms to the rate of ballast degradation. Table 2 presents the morphological changes of some selected particles at some intervals of the number of load cycles. The changes in particle morphological properties increase with an increase in load cycles. However, comparing the rate of changes in particles morphological features, it was observed that the LAA test yielded a higher rate of particles abrasion and breakage than the cyclic loading test. This could be due to the severity of the particles breakage and rate of abrasion under the LAA test in the laboratory being higher than in the field, which agrees with findings of Moaveni et al. (2016).

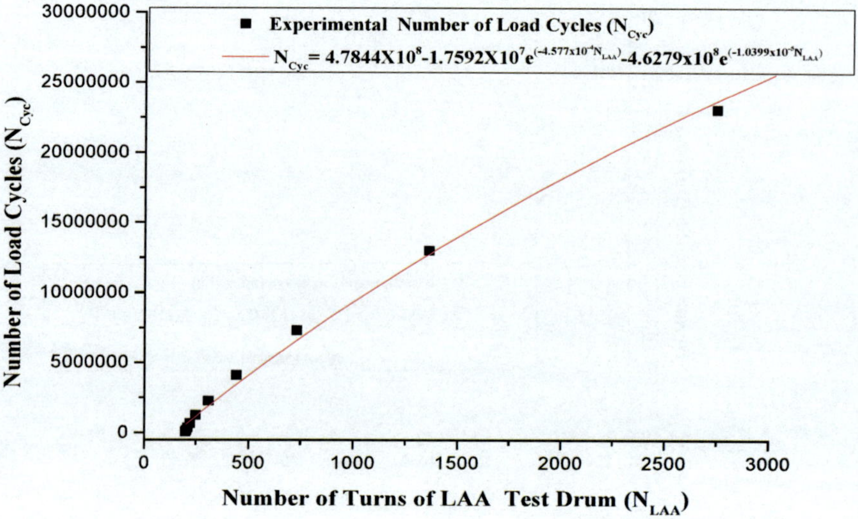

Fig. 10. Relation between number of turns of LAA test drum and the number of load cycles.

4 Conclusion

This study evaluated ballast degradation using the Los Angeles abrasion (LAA) and cyclic loading tests. Changes in the ballast aggregate shape indexes (form, angularity, surface texture, and volume) were also evaluated using image analysis techniques. A typical heavy haul freight train with an average speed of 120 km/h and 30-tons capacity was simulated for the cyclic loading test. At the end of the study, the following conclusions are constituted.

I. The test results revealed that the LAA test and the cyclic loading test yielded different breakage index and changes in morphological properties. It is observed that the degradation indexes under different testing conditions correlate with the rate of changes in the ballast morphological properties.
II. Relationships between ballast breakage index (B_g) with the number of turns of LAA test drum (N_{LAA}) and the number of load cycles (N_{Cyc}) are proposed. The relationships are as presented below;

$$B_g = 17.0233 + 1.2347 N_{LAA} - 32611.2014 N_{LAA}^{-1.5} \tag{7}$$

$$N_{LAA} = 0.9728 e^{0.3921 B_g} \tag{8}$$

$$\mathrm{B_g} = -30.0943 + 1.8286 \times 10^{-13}\left(\mathrm{N_{cyc}^2}\right) + 2.8397 \ln\left(\mathrm{N_{cyc}}\right) \tag{9}$$

$$N_{CYc} = 76381.52 e^{0.2857 B_g} \tag{10}$$

$$\mathrm{N_{Cyc}} = 4.79 \times 10^8 - 1.76 \times 10^7 e^{\left(-4.58 \times 10^{-4} N_{LAA}\right)} - 4.63 \times 10^8 \mathrm{e}^{\left(-1.04 \times 10^{-5} N_{LAA}\right)} \tag{11}$$

These relationships could aid railway practitioners in predicting the number of load cycles at which the railway ballast is due for maintenance.

Table 1. Image-based shape index properties of a single particle through the degradation process based on LA abrasion tests.

Number of Turn of LAA test drum	0	500	1000	1500	2000
Properties	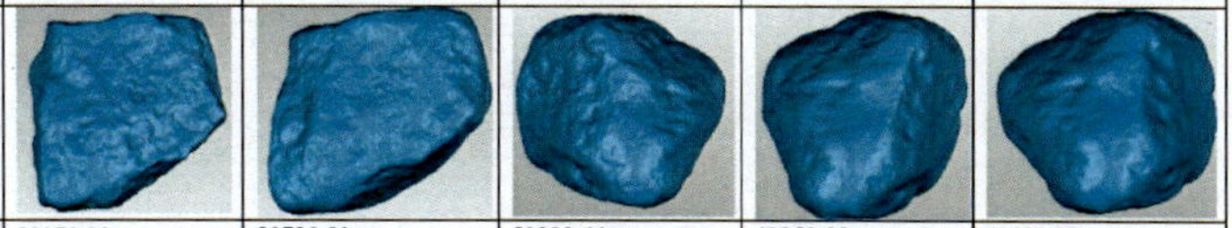				
Volume (mm^3)	51078.03	50730.81	50298.64	49359.29	49033.34
Area (mm^2)	7691.377	7485.89	7359.21	7177.91	7104.52
Angularity index (AI)	2887.70	2517.26	2242.85	2112.30	2036.80
Surface texture Index (SI)	358.59	245.42	225.69	206.30	201.30
Flatness & Elongation Index	1.86	1.84	1.79	1.71	1.54

Table 2. Image-based shape index properties of a single particle through the degradation process based on cyclic loading test

Number of Load Cycles	200,000		400,000		600,000		800,000		1,000,000	
	Before	After	Before	After	Before	After	Before	After	Before	After
Properties	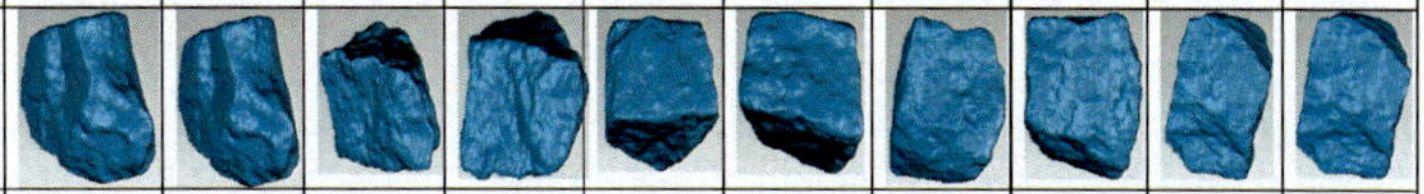									
Volume (mm^3)	41741.56	41626.36	41918.06	41876.98	37571.70	37494.04	35537.90	35180.89	46276.88	45964.24
Area (mm^2)	7014.41	6995.18	7193.29	7189.96	6616.51	6603.45	6670.24	6626.64	7565.30	7426.14
Angularity index (AI)	2995.73	2798.45	2721.55	2302.6	2631.65	2299.56	2312.30	2038.80	2671.57	2333.41
Surface-texture Index (SI)	338.59	322.45	258.79	249.59	248.41	235.69	296.40	271.70	284.90	250.50
Flatness & Elongation Index	1.83	1.83	1.76	1.72	1.89	1.89	1.71	1.68	1.91	1.88

References

ASTM, A.: C535–09 Standard Test Method for Resistance to Degradation of Large Size Coarse Aggregate by Abrasion and Impact in the Los Angeles Machine: Conshohoken, PA: ASTM (2009)

Aursudkij, B.: A laboratory study of railway ballast behaviour under traffic loading and tamping maintenance. University of Nottingham (2007)

Cui, S., Pittman, R.W., Zhao, J.: Restructuring the Chinese Freight Railway: Two Scenarios. Available at SSRN 3229378 (2018)

Danesh, A., Palassi, M., Mirghasemi, A.A.: Evaluating the influence of ballast degradation on its shear behaviour. Int. J. Rail Transp. **6**(3), 145–162 (2018)

Esveld, C., Esveld, C.: Modern railway track, vol. 385. MRT-productions Zaltbommel, Netherlands (2001)

Fischer, S., Németh, A.: special laboratory testing method for evaluation particle breakage of railway ballast material. Sci. Transp. Prog. Bulletin dnipropetrovsk Nat. Univ. Railway Transp. **2**(74), 87–102. https://doi.org/10.15802/stp2018/130854

Guo, Y., Markine, V., Zhang, X., Qiang, W., Jing, G.: Image analysis for morphology, rheology and degradation study of railway ballast: a review. Transp. Geotech. **18**, 173–211 (2019)

Han, X.: The role of particle breakage on the permanent deformation of ballast (2012)

Indraratna, B., Salim, W.: Modelling of particle breakage of coarse aggregates incorporating strength and dilatancy (2002)

Indraratna, B., Nimbalkar, S., Rujikiatkamjorn, C.: From theory to practice in track geomechanics–Australian perspective for synthetic inclusions. Transp. Geotech. **1**(4), 171–187 (2014)

Indraratna, B., Nimbalkar, S., Coop, M., Sloan, S.W.: A constitutive model for coal-fouled ballast capturing the effects of particle degradation. Comput. Geotech. **61**, 96–107 (2014). https://doi.org/10.1016/j.compgeo.2014.05.003

Indraratna, B., Sun, Y., Nimbalkar, S.: Laboratory assessment of the role of particle size distribution on the deformation and degradation of ballast under cyclic loading. J. Geotech. Geoenviron. Eng. **142**(7), 04016016 (2016)

Indraratna, B., Tennakoon, N.C., Nimbalkar, S.S., Rujikiatkamjorn, C.: Behaviour of clay-fouled ballast under drained triaxial testing (2013)

Koohmishi, M., Palassi, M.: Degradation of railway ballast under compressive loads considering particles rearrangement. Int. J. Pavement Eng. 1–13 (2018)

Lu, M., McDowell, G.: Discrete element modelling of railway ballast under monotonic and cyclic triaxial loading. Géotechnique **60**(6), 459 (2010)

Marsal, R.J.: Large-scale testing of rockfill materials. J. Soil Mech. Found. Div. **93**(2), 27–43 (1967)

Masad, E., Fletcher, T.: Aggregate imaging system (AIMS): Basics and applications (2005)

McDowell, G.R., Lim, W.L., Collop, A.C., Armitage, R., Thom, N.H.: Laboratory simulation of train loading and tamping on ballast. Paper presented at the Proceedings of the institution of civil engineers-transport (2005)

Moaveni, M., Qian, Y., Qamhia, I.I., Tutumluer, E., Basye, C., Li, D.: Morphological characterization of railroad ballast degradation trends in the field and laboratory. Transp. Res. Rec. **2545**(1), 89–99 (2016)

Nålsund, R., Tutumuler, E., Horvli, I.: Degradation of railway ballast through large scale triaxial and full scale rail track model tests. In: Paper presented at the Proceedings of the international conferences on the bearing capacity of roads, railways and airfields (2013)

Qian, Y., Boler, H., Moaveni, M., Tutumluer, E., Hashash, Y.M., Ghaboussi, J.: Characterizing ballast degradation through Los Angeles abrasion test and image analysis. Transp. Res. Rec. **2448**(1), 142–151 (2014)

Qian, Y., Boler, H., Moaveni, M., Tutumluer, E., Hashash, Y.M., Ghaboussi, J.: Degradation-related changes in ballast gradation and aggregate particle morphology. J. Geotech. Geoenviron. Eng. **143**(8), 04017032 (2017)

Röthlisberger, F., Cuénoud, J., Chastan, L., Däppen, J., Kuerzen, E.: Compressive Strength of Aggregates on the Stack, p. 59 (2006)

Sun, Q., Indraratna, B., Nimbalkar, S.: Effect of cyclic loading frequency on the permanent deformation and degradation of railway ballast. Géotechnique **64**(9), 746–751 (2014)

TB/T2328.14–2008: Test method for railway ballast. NATIONAL STANDARD OF THE PEOPLE'S REPUBLIC OF CHINA (2008)

Design and Analyses of the Hurricane Protection Floodwall in South Louisiana

Wenjun Dong(✉) and Robert Bittner

Bittner-Shen Consulting Engineers, Inc., 921 SW Washington Street, Suite 765, Portland, OR 97205, USA
wjd@bittner-shen.com

Abstract. This paper presents a comprehensive review on the finite element analyses for the performance-based design of the Floodwall, an important component of hurricane storm surge barrier for New Orleans, Louisiana. The finite element models are calibrated and verified with a lateral pile load test, and they are used to derive the design hurricane storm surging loads on the Floodwall. Such storm surging loads are used to determine the seepage cut-off depth of the wall based on design seepage criteria. The settlement of the Floodwall during its design life is estimated by the finite element consolidation analysis with considering soil-pile-interaction. Finally, a brief discussion is given to the finite element stability analysis for an effective remediation design of the submerged rock dike of the Floodwall using a soil cut-off structures at the dike toe.

1 Introduction

In 2005, Hurricane Katrina hit the city of New Orleans, Louisiana, and the storm surge overwhelmed the existing levee system of the Inner Harbor Navigation Canal (IHNC), resulting in a devastated flooding of the city (Duncan et al. 2008; IPET 2007). In order to protect the city from future hurricanes, the U.S. Army Corps of Engineers decided to build a storm surge barrier across Lake Borgne on the east of New Orleans between the Mississippi River-Gulf Outlet (MRGO) and the Gulf Intercoastal Water Way (GIWW). An important component of the surge barrier is the 3.2-km long Floodwall (Huntsman 2011; Reid 2013). The general soil profile at the construction site consists of Holocene age mash and soft clay overlying Pleistocene deposits. Due to the deep soft ground condition, the Floodwall was designed as a laterally braced piled wall, with a submerged rock dike over the MRGO channel. The high storm surge loading combined with the very deep soft soils created significant design challenges for the Floodwall. These design challenges include determining the appropriate hurricane storm surge loads on the wall, estimating the settlement of the wall during its design life, and evaluating the global stability of the wall under design storm surge loading. A safe and cost-effective design of the Floodwall "floating" in soft-medium-stiff clays required a thorough understanding of the soil-pile-interaction to satisfy design requirements for structural strength and serviceability. For this purpose, finite element Soil-Structure-Interaction (SSI) analysis models were developed to assess the performance of the Floodwall under the design hurricane surge condition (Dong and Schwanz 2011; Dong 2016).

© The Author(s), under exclusive license to Springer Nature Switzerland AG 2021
S. Shu et al. (Eds.): GeoChina 2021, SUCI, pp. 59–70, 2021.
https://doi.org/10.1007/978-3-030-80142-7_5

This paper presents a comprehensive review on finite element model analyses that were used to assess the performance of the Floodwall under storm surging loads, with respect to several important geotechnical aspects, including soil-pile interaction, seepage, settlement and rock dike stability. First, the general condition of the subsurface profile at the site and the details of the Floodwall are presented. Next, the finite element SSI model are briefly described. Then, the calibration and verification of the soil input parameters and finite element model with the results of a lateral pile load test are discussed. The finite element SSI seepage analyses are used for deriving the appropriate hurricane storm surge loads on the wall, which are used to determine the seepage cut-off of the wall. Next, the finite element settlement analysis that is used to estimate the consolidation of the Floodwall over the MRGO channel with a submerged rock berm is described. Finally, a brief discussion is given to the finite element stability analysis for the effective remediation design of the submerged rock dike of the Floodwall using a soil cut-off structures at the toe.

2 General Subsurface Profile

The general subsurface profile at the construction site include a very soft organic surficial clay (or Marsh) layer underlain by the Interdistributary and Prodelta soft clay deposits. The soft clay is underlain by Pleistocene medium and stiff clay, with an intrusion layer of nearshore gulf (Interdelta) sand in some areas. Both soil borings and field Cone Penetrometer Tests (CPT) were made at several locations to determine the soil profile and shear strengths along the structure alignment. The soil profile and design soil shear strengths for a typical Floodwall design ranch are shown in Table 1 (EUSTIS 2009). In Table 1, γ_{sat}, s_u, ϕ, $Nu_u k_h$ and k_v represent the saturated unit weight, shear strength, internal friction angle, undrained Poisson ratio, horizontal and vertical hydraulic conductivities of soil layers. The soil hydraulic conductivities were estimated based on the design manual DIVR-1110-1-400 (USACE 1998).

Table 1. General soil profile and design parameters

Elevation (m)	Soil Type (USCS)	γ_{sat} (kN/m^3)	s_u (kPa)	Φ (deg)	ν_u	k_h/k_v (10^{-3} m/day)
0 to −3.7	PT	10.8	4.8		0.4	1.7/0.5
−3.7 to −6.1	CH	16.5	5.0		0.4	1.7/0.5
−6.1 to −15.2	CH	15.7	5.0 –21.7		0.4	1.7/0.5
−15.2 to −18.3	CH	17.3	25.2 –32.3		0.4	1.7/0.5
−18.3 to −21.3	SM	19.7		30	0.4	73.5/37.0
−21.3 to −35.1	CH	18.1	37.3–60.0		0.42	1.7/0.5
−35.1 to −51.8	CL	17.4	52.3–91.9		0.45	1.7/0.5
Below −51.8	CL	17.4	52.3–103.9		0.45	1.7/0.5

3 Floodwall

The Floodwall is built over the Lake Borgne and MRGO at the east of New Orleans, South Louisiana, with a total length of about 3.2 km. Along with a sector gate over the GIWW, the Floodwall creates a surge barrier to prevent future storm surges from entering the City of New Orleans during hurricane seasons. The typical Floodwall is composed of vertical precast post-tensioned cylindrical concrete piles with an outside diameter of 1.7 m and a thickness of 15.2 cm. These large diameter cylindrical concrete piles are driven closely and braced by steel-pipe batter piles with a batter angle of 34° (or a slope of 1 H:1.5 V). The embedment length of the plumb piles of the Floodwall is about 35 m from the mudline at EL-4.6 m, and their center-to-center spacing is 1.8 m. The steel batter piles have a wall thickness of 1.9 cm above EL-8.2 m and 1.3 cm below and were driven into the ground to EL-58 m with a center-to-center spacing of 3.6 m. The interstitial space between plumb cylindrical piles of the Floodwall was closed by the irregular pentagonal closure piles with jet grout applied to seal the gap between the closure and cylindrical piles. The Floodwall provides a flood protection to El + 7.6 m with wave deflectors to El + 7.9 m. Figure 1 shows a photograph of the Floodwall during construction. The soft clay foundation conditions and large storm surge loading combined to create design challenges of the Floodwall. A safe and cost-effective design of the Floodwall "floating" in soft-medium-stiff clayey soils required a thorough understanding of soil-pile-interaction to satisfy design requirements for structural strength and serviceability. For this purpose, soil-structure-interaction finite element model analyses were conducted to assess the performance of the Floodwall under design hurricane surge loading conditions.

Fig. 1. Floodwall with batter piles during construction

4 Finite Element Model

4.1 Model Configurations

The finite element SSI model of the Floodwall were developed using a commercial geotechnical computational program PLAXIS 2D (Brinkgreve 2002). In the 2D finite element model, the elasto-plastic plate (or beam) elements were used for both vertical and batter piles. The structural properties (e.g. axial stiffness and flexural rigidity) of elements were determined by dividing pile properties by the pile center-to-center spacing. This smearing approach was also applied to define unit weights of beam elements. The connection of the vertical pile to the cap beam is fixed, but a pinned connection was used for the batter pile to cap beam. The advanced hardening soil model was used in the finite element analyses to describe the nonlinear soil reaction behavior to pile deformation (Brinkgreve 2002; Duncan and Chang 1970). The computational domain was discretized using high order (6-node) triangular elements. To save the computational time, a finer mesh was used in the soil wedges adjacent to vertical piles while the coarser mesh was used in the region far from the Floodwall structure. Sensitivity analyses have been performed to verify the accuracy of such selected finite element mesh. Based on the soil shear strength characteristics, an average strength reduction factor of 0.67 was applied to the soil-structure interface elements to simulate the soil adhesion to piles. The standard fixity boundary conditions were used in the finite element model, i.e. fixed in horizontal direction on the vertical boundaries and fixed in both horizontal and vertical directions on the bottom of the model. A close boundary condition was also set on the bottom of the model during the groundwater seepage calculations. Figure 2 shows the finite element model configuration. The calculation phases were created in the finite element modeling for major construction sequences such as construction of the dredged channel, installation of piles and the placement of scour stone layer. Additional calculation phases were also created for different design loads cases, including hurricane loading case. Based on the evaluation of geological and hurricane condition, the long-period surging load to the top of wall is considered critical for the design of the Floodwall, and it is modeled as the hydrostatic water load in the finite element analyses.

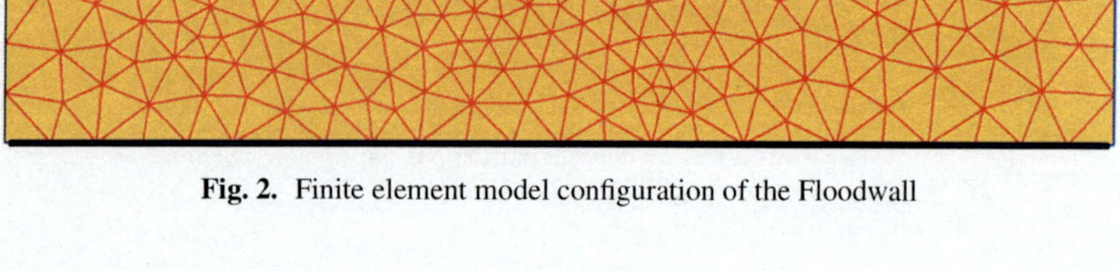

Fig. 2. Finite element model configuration of the Floodwall

4.2 Lateral Pile Load Test

To better model the soil-pile-interaction behavior in the 2D finite element SSI analysis, the lateral resistance of local soft clays to laterally loaded large-diameter piles were calibrated by the results of a lateral pile load test. The lateral pile load test was performed on a

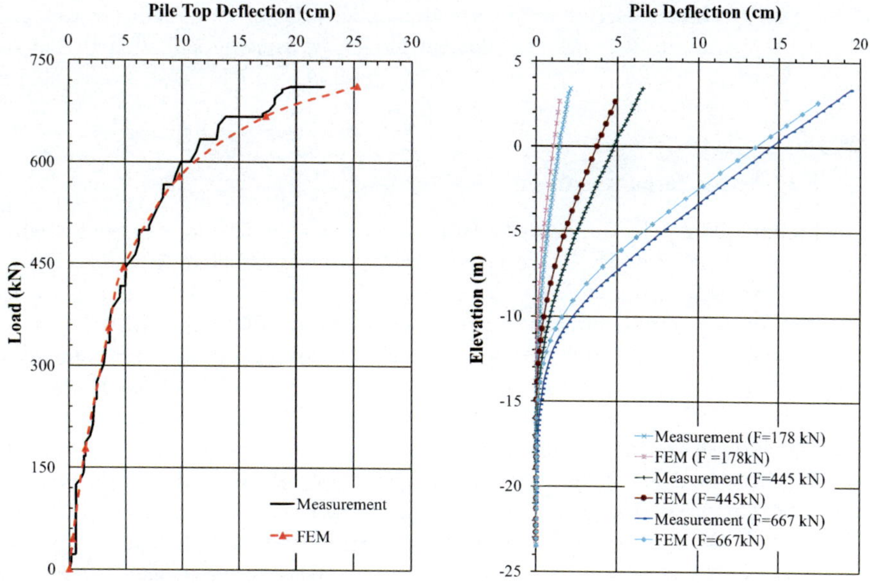

(a) Pile top deflections (b) Pile deflections

Fig. 3. Field measurements and numerical predictions of load-deflection behavior of the test pile

prototype cylindrical precast post-tensioned concrete pile of the Floodwall (Reese 2009). Using the revised Matlock approach (Matlock 1970; Lee and Gilbert 1979; Dong and Schwanz 2011), the in-situ stiffness's of soft clay under the single pile deformation were derived and they are used in 2D finite element models to reflect 3D pile-soil-interaction. Figure 3 shows the comparison of the numerical predictions and field measurements on the deformation of tested pile under different test loads. The good agreement of numerical predictions with field measurements supports the reconciliation of soil stiffness's with the *p-y* interpretation from the lateral pile load test and the rigorousness of the finite element modelling for the soil-pile interaction.

5 Soil Structure Interaction

5.1 Pile-Soil Interaction

The plumb piles of the Floodwall were installed closely, resulting in the overlapping of the activated passive soil wedges of laterally loaded adjacent piles. Such overlapping would result in an increase in the soil stress level within the interfered passive wedges of the pile, and consequently leading to an increase in soil strain based on the stress-strain relationship of the soils. In order to take into account this pile group interaction effect, soil wedges from pile tips to mudline are introduced in the 2D finite element models in the vicinity of the plumb pile wall (see Fig. 1). Using the soil wedge concept, the group effect of closely spaced vertical cylinder piles of the Floodwall was taken into consideration in the soil finite element model by reducing soil strength and stiffness parameters by a group reduction factor (Ashour et al. 1998; 2004; Dong and Schwanz 2011).

5.2 Active Pore Pressures

First, the finite element model was used to investigate the storm surging loads on the Floodwall under design hurricane conditions, which would induce a high-water level at EL 7.5 m at flood side and a low-water level at EL 0.5 m at the protected side, respectively. The corresponding hydrostatic water pressures on the wall to this high and low water level are shown Fig. 4(a). Both steady-state and transient seepage analyses were performed using soil hydraulic conductivity as shown in Table 1, and corresponding steady-state and transient pore water pressures on the Floodwall in the soil layers are also shown in Fig. 4(a). The steady-state pore pressures correspond to the long-term ultimate stage of the subsurface flow under storm surging condition, which would only be achieved in a relatively long period in the clayey soil materials. The transient pore water pressures, on the other hand, correspond to the short-term subsurface flow state under storm surging condition, which only takes place in the shallow soil layers during one-day surging period due to low permeability of soils. However, both long-term steady state and short-term transient pore water pressures do not include the water surcharge effect on foundation soft soils under the storm surging condition. During one-day storm surging period, the foundation clay layers can be considered to have undrained behaviors, and therefore, excess pore water pressures would develop under the water volume surcharge on the flood side. Such excess pore pressures also include the effect of soil volume change due to the SSI effects of the Floodwall with adjacent soils. Thus, the active pore pressures on the Floodwall under storm surge loading condition consists of the transient pore pressures and the excess pore pressures due to the undrained behavior of the clay layers (Dong, 2016). Such active pore pressures are captured by the finite element SSI analysis, and they are shown in Fig. 4(a) as well.

5.3 Model Results

With the verification of the finite element model with lateral pile load tests and comprehension on storm surging loads on the wall, the finite element SSI models were used to assess the performance of the Floodwall under design hurricane loading condition. Figure 4(b) shows the computational results of the unfactored bending moment distribution along the vertical pile corresponding to critical design load case including high storm surge water pressures on the flood side. Figure 4(b) also includes numerical results from a *p-y* curve-based GROUP model analyses (Reese et al. 2010). In GROUP model analyses, the storm surge induced net unbalanced pressures on the vertical pile wall below the mudline were derived from the finite element SSI analysis. A very good agreement of the design bending forces on the vertical pile was observed between the finite element analysis and GROUP model as shown in Fig. 4(b). The discrepancy in the numerical results in pile bending moment below EL-15 m mainly attributes to the fact that the finite element model analysis indicate the soil movements in the vicinity of piles, which would release the restrains on piles and reduce the bending demands on piles. Such soil movements were not included in the GROUP analysis.

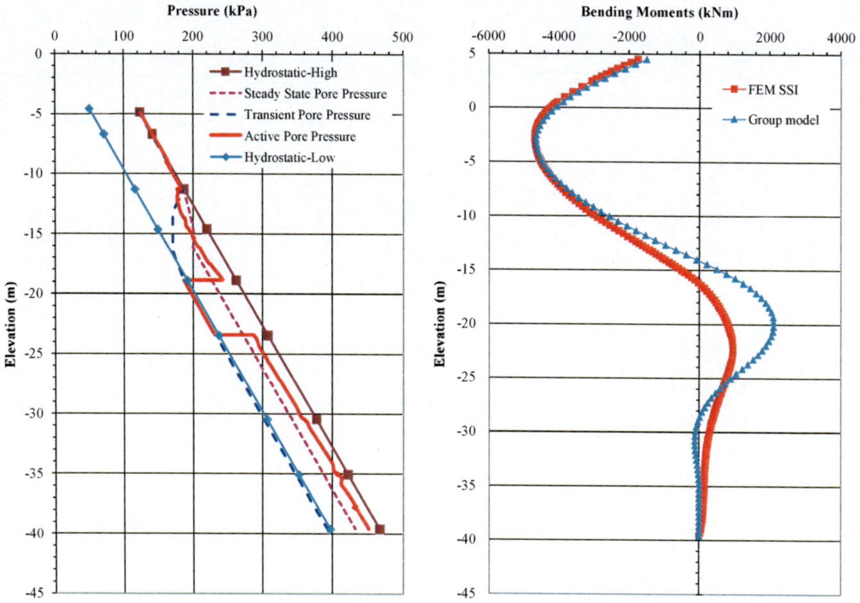

(a) Pressure distributions (b) Bending moments of plumb piles

Fig. 4. Outputs of finite element SSI analyses of the Floodwall

6 Seepage Cutoff

As stated previously, closure piles with grout soil columns were installed between the cylindrical piles to provide a seal to reduce leakage. The depth of the grouted soil column for seepage cutoff was determined based on three seepage criteria as required by Hurricane and Storm Risk Reduction System Design Guide (HSDRRS) (USACE 2012) including exit gradient, Terzaghi piping and heave criteria (Terzaghi 1995; Dong 2012). The exit gradient criterion requires that the critical hydraulic gradient of subsurface flows be greater than the exit gradient, to avoid zero effective stress condition. Terzaghi piping criterion requires that the buoyant weight of overlying soils be greater than the upward seepage force at seepage cut-off tip, to avoid an effective heave phenomenon that a mass of soil may be lifted initially then followed by piping. The heave criterion is like Terzaghi piping criterion but is based on total stress approach. Using these design criteria and the active pore water pressures obtained from finite element SSI analyses, as presented in Fig. 3(a), the distributions of safety factors for seepage cut-off along the Floodwall in one design reach per different design criteria are calculated and shown in Fig. 5. The dashed red lines in Fig. 5 represent the minimum factors of safety required by the HSDRRS for each seepage criterion. Based on these minimum factors of safety, the required seepage cut-off depth of the Floodwall for each criterion can be easily obtained from these safety factor distributions.

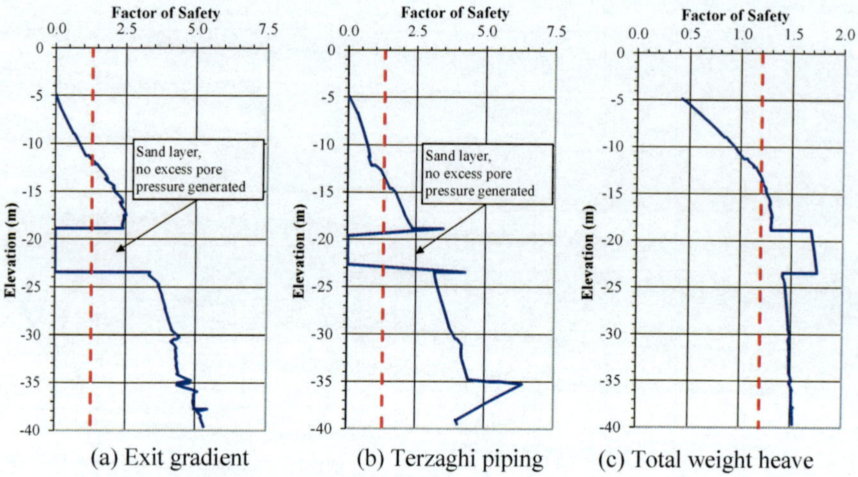

Fig. 5. Distributions of safety factors based on design seepage cutoff criteria

7 Settlement Analyses

7.1 Consolidation Model

The southern portion of the Floodwall was built across the MRGO channel that has a maximum depth of about 12.8 m below mean sea level at the EL + 0.0. This portion of Floodwall is referred as the MRGO Closure, and it requires a submerged rock berm with a top elevation to EL-4.5 m over the outlet channel to provide lateral supports to the wall. The berm consists of two rock dikes on both sides of the wall with a side slope of 1 V:9 H and a sand core in between to facilitate driving of the piles. A final 6-feet rip rap stone layer with a side slope of 1 V:3 H was placed atop of the sand core for scour protection. The max length of the berms along the channel at top and bottom are about 49 and 175 m, respectively. The placement of the rock berm with a maximum thickness of about 8.2 m induces a significant consolidation of the soft soils beneath. Such settlement had to be addressed in the design of the MRGO Closure. For this purpose, a finite element soft soil creep (SSC) model (Brinkgreve 2002) was created to estimate the consolidation of the MRGO Closure during its 50-year design life. Figure 6 shows the cross section of the MRGO Closure at the center of outlet channel with a mudline at EL-12.8 m.

7.2 Consolidation Calculations

First, the consolidation of the rock berm along the central line of the outlet channel (with a thickness of about 8.2 m) with a 50-year design life was investigated using

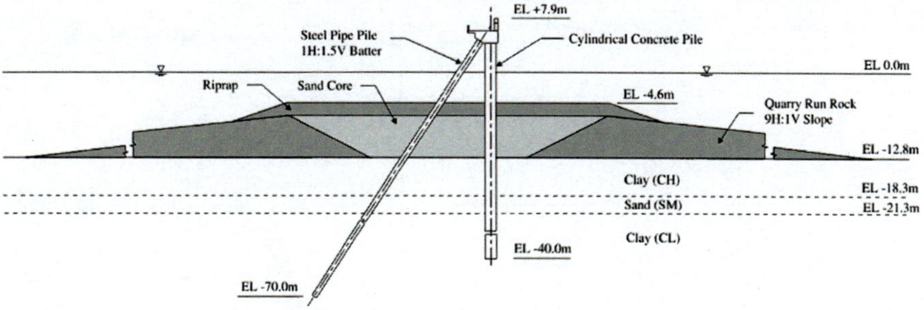

Fig. 6. Cross section of the MRGO Closure with rock berm at the outlet channel center (mudline at EL-12.8 m)

the 2D finite element SSC model. The corresponding settlement profiles corresponding to different time spans, obtained from the finite element consolidation analyses, are shown in Fig. 7(a). These numerical predictions indicated a settlement of about 45 cm for the MRGO berm along its center line in 50 years. However, these consolidation calculations are only for the rock berm and they do not consider the influence of the Floodwall structures built through it. To estimate the settlement of the MRGO Closure during its design life, the finite element model was rerun again with the Floodwall pile structures. The equivalent structural properties of the vertical and batter piles in the 2D finite element model were determined by considering the spacing of vertical pile and batter pile. The unit weight of vertical pile includes the weight of closure piles, jet grout around pile and buoyance effects. The corresponding settlement profile along the center line of the berm, obtained from the finite element analyses, is shown in Fig. 7(b). It can be seen by comparing Figs. 7(a) and 7(b) that the less consolidation of the berm can be expected due to the interaction between pile and soils. With the pile structures installed, the adjacent settling soil under surcharge will create down-drag forces on the piles, and thus the surcharge loads are transferred to deeper more competent soils. The settlement of the MRGO Closure in 50 years is approximately equal to the soil settlement of about 20 cm at the vertical pile tip elevation of EL-40 m (see Fig. 7(b)).

8 Stability of Rock Berm

8.1 Deep-Seated Soil Failure Below Rock Dike.

During the construction of the rock berm of the MRGO Closure, field observations indicated rotational displacements of the rock berm and implied a global slip movement of the soft soils beneath the rock dike on the flood side. Bathymetric surveys identified mud waves at toe of the rock dike on the flood side, accompanying a displacement of the sand core and the upper portion of rock dike. This deep-seated soil failure beneath the rock dike was successfully captured by the finite element stability analysis (Dong et al. 2012). The ϕ-c reduction approach was used in the finite element stability analysis.

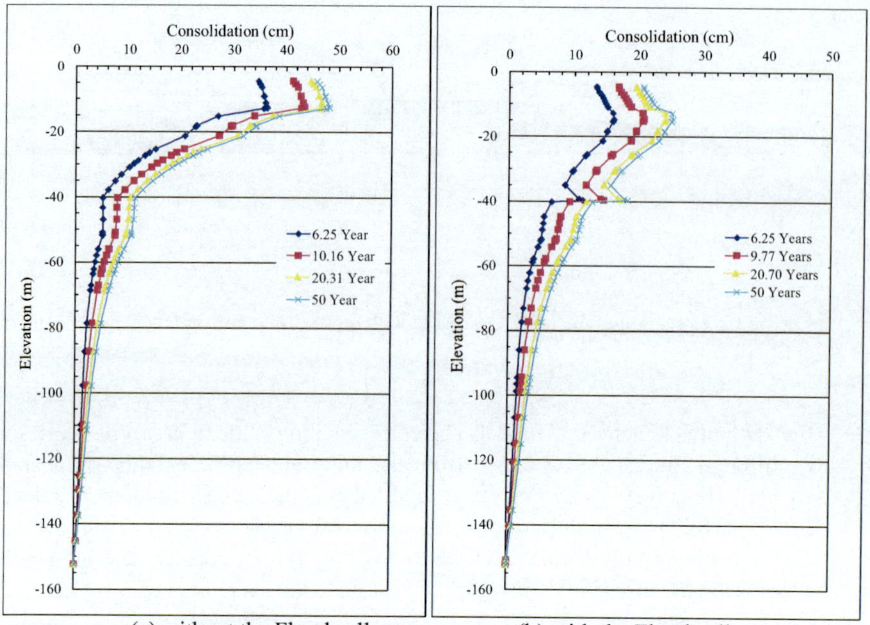

(a) without the Floodwall (b) with the Floodwall

Fig. 7. Finite element settlement estimates of the MRGO Berm

The analysis showed a safety factor of 1.1 for the as-built condition of the rock dike, indicating a marginally stable conditions of the rock dike as observed. This deep-seated slip failure in the weak soil layer created displacement of dick crest and a mud wave at the rock dike toe as shown in Fig. 8(a).

8.2 Soil Cut-Off Structures

In order to increase the stability of the rock berm, the rock dike on the flood side was redesigned with a sheet-pile soil cut-off structure at its toe. The penetration depth and grade of the sheet-piles were determined based on the finite element SSI stability analysis with a minimum target safety factor of 1.5. Figure 8(b) shows the corresponding critical failure surface with a safety factor equal to 1.51, obtained from finite stability analysis with a sheet pile embedment of 13 m. With such embedment length, the sheet pile cut-off structure penetrated through the weak soil layer and embedded into the stronger soils below. It can be seen from the Fig. 8(b) that the slip failure surface and mud wave that occurred previously in the weak soil layer were successfully cut-off by the sheet pile wall.

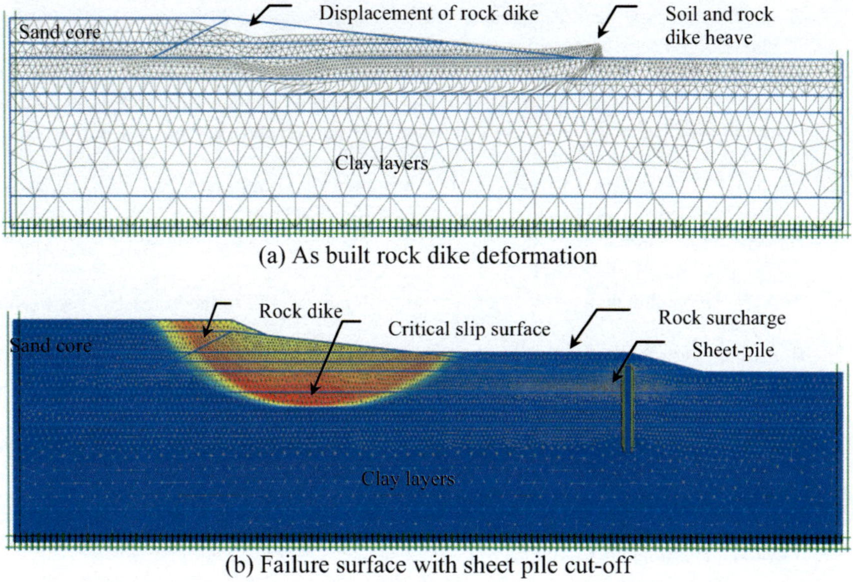

Fig. 8. Finite element stability analysis results of rock dike

9 Conclusions

Finite element model analyses have been successfully used for the performance-based design of the Floodwall that was built in the soft soil foundation to satisfy the design requirements for hurricane storm surge barrier system. The soil input parameters and the finite element models are calibrated and verified by the results of a lateral pile load test. The finite element SSI analyses are used to address several important design challenges including determining appropriate hurricane storm surge loads on the wall and the seepage cut-off depth of the wall, predicting the settlement of the wall during its design life, and the effective remediation design of the rock berm of the MRGO Closure. With the aid of the finite element modeling analyses, the Floodwall was designed safely and economically.

References

Ashour, M., Norris, G.M., Pilling, P.: Lateral loading of a pile in layered soil using the strain wedge model. J. Geotech. Geoenviron. Eng. **124**(4), 303–315 (1998)

Ashour, M., Pilling, P., Norris, G.M.: Lateral behavior of pile group in layered soils. J. Geotech. Geoenviron. Eng. **130**(6), 580–592 (2004)

Brinkgreve, R.B.J.: PLAXIS Manual. Delft University of Technology & PLAXIS b.v, The Netherlands (2002)

Dong, W., Schwanz, N.: Finite element soil-pile interaction analysis of Floodwall in soft clay. In: Han, J., Alzamora, D.E. (eds.) Proceedings of Geo-Frontiers 2011-Advances in Geotechnical Engineering, Dallas, Texas, pp. 4378–4387 (2011)

Dong, W.: Finite element seepage analysis of the floodwall in South Louisiana.In: Proceedings of Ports 2016: Port Engineering, New Orleans, Louisiana, pp. 223–232 (2016)

Dong, W., Yao, S., Hill, A., Neil, S.: Investigations of stability improvements for submerged embankment on soft clay in South Louisiana. In: Proceedings of ASCE GeoCongress 2012-State of the Art and Practice in Geotechnical Engineering, Oakland, California, pp. 624–633 (2012)

Duncan, J.M., Brandon, T.L., Wright, S.G., Vroman, N.: Stability of I-walls in new orleans during Hurricane katrina. J. Geotech. Geoenviorn. Eng. **134**(5), 681–691 (2008)

Duncan, M., Chang, C.Y.: Nonlinear analysis of stress and strain in soils. J. Soil Mech. Found. ASCE **96**(5), 1629–1653 (1970)

EUSTIS: Geotechnical Interpretation Report (GIR). EUSTIS Engineering LLC, New Orleans, LA (2009)

Huntsman, S.R.: Design and construction of the lake borgne surge barrier in response to Hurricane karina. In: Magoon, O.T., Noble, R.M., Treadwell, D.D., Kim, Y.C. (eds.) Proceedings of the 2011 Conference on Coastal Engineering Practice, pp. 117–130 (2011)

IPET. Performance Evaluation of the New Orleans and Southeast Louisiana Hurricane Protection System. Final Report of the Interagency Performance Evaluation Taskforce, US Army Corps of Engineers (2007)

Lee, P.Y., Gilbert L.W.: Behavior of laterally loaded pile in very soft clay. In: Proceedings 11th Annual Offshore Technology Conference, Houston, pp. 387–395 (1979)

Matlock, H.: Correlations for design of laterally loaded piles in soft clay. In: Proceedings Second Annual Offshore Technology Conference, Houston, pp. 577–594 (1970)

Reese, L.C., et al.: Report of Lateral Pile Load Test in Primary Floodwall System at Inner Harbor Navigation Canal. Orleans and St. Bernard Parishes, Louisiana (2009)

Reese, L.C., Wang, S.-T., Vasquez, J.: A Program for the Analysis of a Group of Piles Subjected to Vertical and Lateral Loading. Version 8, Ensoft, Inc., Austin, Texas (2010)

Reid, R.L.: Defending New Orleans, special report. Civil Eng. Mag. ASCE **83**(11), 48–67 (2013)

Terzaghi, K., Peck, R.B., Mesri, G.: Soil Mechanics in Engineering Practice-, 3rd edn. John Wiley & Sons, New York (1995)

U.S. Army Corps of Engineers. Hurricane and Storm Damage Risk Reduction System Design Guidelines, Section 3.0 – Geotechnical (Updated 12 Jun 2008), New Orleans District Engineering Division, New Orleans, LA (2012)

U.S. Army Corps of Engineers. Soil Mechanics Design Data, Section 8 – Ground Water and Seepage, DIVR 1110–1–400 Engineering and Design, Mississippi Valley Division Vicksburg, Mississippi (1998)

Study on Stability of Tunnel Portal Section Based on Strength Reduction Shortest Path Method

Wei Wang, Guiqiang Gao(✉), Mingjun Hu, Yanfei Zhang, and Haojie Tao

School of Civil Engineering, Central South University, Changsha 410075, China
albertgao@csu.edu.cn

Abstract. It is very significant to maintain the stability of surrounding rock during the entire lifetime of the tunnel. Generally speaking, the stability of the surrounding rock in the portal section of the tunnel is the worst. Therefore, it is meaningful to study the instability mechanism and factors affecting the instability of the surrounding rock in the tunnel portal section. In this paper, we studied the stability of tunnel surrounding rock based on a novel double strength reduction method (the strength reduction shortest path method). The double strength reduction method was used to reduce the cohesion and internal friction angle with different reduction factors in the stability analysis of surrounding rock. We used the overall safety factor to evaluate the stability of the tunnel. The smaller the overall safety factor, the more likely the tunnel is to lose stability. The influence of strength parameters of surrounding rock on the stability of the portal section was discussed emphatically. The results show that: As the reduction ratio λ increases, the length of the strength reduction path first decreases and then increases, which indicates that there is a shortest strength reduction path; It is reasonable to take the drastic change of the vault displacement as the instability criterion; As the initial internal friction angle and the initial cohesion increase, the length of the strength reduction shortest path and the overall safety factor increase, in which the effect of the internal friction angle is more significant.

This paper can provide a new method for studying the stability of the tunnel portal section and provide a useful reference for future tunnel construction.

Keywords: Shallow buried mountain tunnel · Tunnel portal section · Surrounding rock stability · Double strength reduction method · The shortest strength reduction path method

1 Introduction

The surrounding rock of the tunnel portal section is usually severely weathered. The rock mass there is loose and broken, which probably leads to slope instability. Therefore, it is of great practical significance to analyze the instability mechanism, and factors affecting the instability of the surrounding rock in the tunnel portal section. At present, the commonly used methods for stability analysis of rock and soil mass include the limit

© The Author(s), under exclusive license to Springer Nature Switzerland AG 2021
S. Shu et al. (Eds.): GeoChina 2021, SUCI, pp. 71–91, 2021.
https://doi.org/10.1007/978-3-030-80142-7_6

equilibrium method [1, 2], limit analysis method [3, 4], and strength reduction method [5]. Among all the methods, when using the strength reduction method, the safety factor and failure position of the slope can be obtained without assuming the location and form of the slip plane. And large-scale finite element software such as ABAQUS, Midas can provide convenience for the solution. Consequently, this method is widely used in engineering and is the dominating method to analyze the stability of complex rock and soil mass.

The reduction mechanism of the traditional strength reduction method is to reduce the cohesion and internal friction angle in the same proportion. By increasing the reduction coefficient, the overall safety factor can be obtained when the rock and soil reach the critical instability state [6, 7]. However, this method ignores that the cohesion and internal friction angle have a different impact on the stability of rock and soil. To make up for the defect of the traditional strength reduction method, the double strength reduction method that uses different reduction factors was proposed [8–11]. When the double strength reduction method is used to study the stability of tunnels, the uncertainty of tunnel instability modes will reduce the applicability of commonly used instability criteria. Therefore, it is necessary to focus on determining an instability criterion suitable for tunnel stability analysis. Besides, determining the appropriate reduction coefficient is the key and basis for the application of this method. There are few studies on the stability of the tunnel portal section based on the double strength reduction method at present. And most relative studies used two-dimensional models, which couldn't reflect the engineering geological conditions and the distribution characteristics of the free face. The three-dimensional model based on the strength reduction method is usually used in slope stability research [12, 13].

In this paper, based on a novel double strength reduction method (the strength reduction shortest path method), a three-dimensional model of the tunnel portal section was established to reveal the instability law of rock and soil mass. The research results of this paper could provide a reference for scientific researches and engineering application of tunnel engineering.

2 Strength Reduction Shortest Path Method

2.1 Reduction Mechanism

To obtain reasonable reduction factors, ISAKOV et al. [14] proposed a new method, namely the strength reduction shortest path method, in a study on the subgrade stability in 2010. Then they successfully applied this method to the slope stability analysis in 2014 [15]. They reduced the cohesion c and the internal friction angle φ according to different reduction factors. The reduction factor of cohesion was F_c, and the reduction factor of internal friction angle was F_φ, as shown in formula (1). The definition of the reduction factors in this method is the same as that of the double strength reduction method.

$$\begin{cases} F_c = \frac{c_0}{c_{SRT}} \\ F_\phi = \frac{tan\phi_0}{tan\phi_{SRT}} \end{cases} \tag{1}$$

Where c_0 and φ_0 are the initial cohesion and the initial internal friction angle, respectively, c_{SRT} and φ_{SRT} are the reduced cohesion and the reduced internal friction angle, respectively.

The ratio of F_c to F_φ is defined as the parameter reduction ratio λ, namely $\lambda = F_c / F_\varphi$. In Fig. 1, M_0 represents the original state of the tunnel, M_n represents a certain limit state of the tunnel (n is a positive integer). The reduction path L can be obtained by connecting M_0 and M_n, and its expression is show in formula (2).

$$L = \sqrt{\left(1 - \frac{1}{F_C}\right)^2 + \left(1 - \frac{1}{F_\varphi}\right)^2} \tag{2}$$

The reduction ratio λ theoretically has innumerable values, and correspondingly there are countless reduction paths. The limit state line is obtained by connecting all the limit state points. L_k is the traditional strength reduction path, and its intersection point with the limit state line is M_k. Among all reduction paths, there must be a reduction path that is the shortest, which is expressed as L_{min} in Fig. 1. Through this shortest reduction path (the shortest path of strength reduction), the tunnel can reach the limit state at the fastest from the initial state.

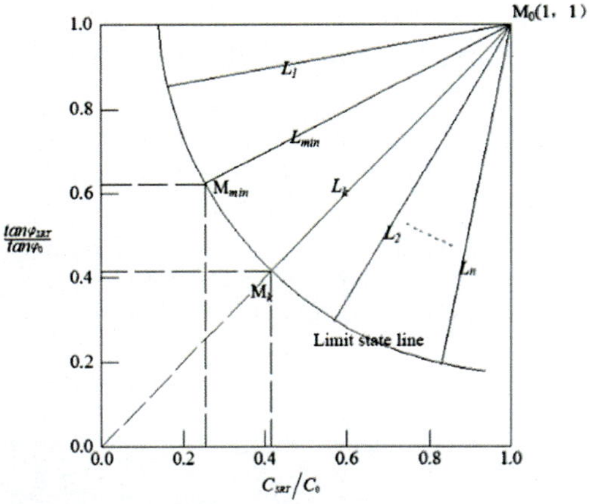

Fig. 1. Schematic diagram of the limit state line

Pan Jiazheng proposed the principle of maximum and minimum stability of rock and soil mass [16]. He believes that when the rock and soil mass become unstable, the sliding surface will provide the maximum anti-sliding force it can provide. However, this maximum anti-sliding force is the smallest of all potential sliding surfaces. Pan's principle is similar to the strength reduction shortest path method and can help us get a physical interpretation of the strength reduction method.

2.2 Definition of the Overall Safety Factor

When calculating the overall safety factor of the shear failure of the tunnel surrounding rock based on the traditional strength reduction method, because the cohesion and the internal friction angle are reduced in the same proportion, the overall safety factor is $F_s = F_c = F_\varphi$. In this case, the overall safety factor has a clear physical meaning and can be solved simply. However, when adopting the double strength reduction method, it is difficult to calculate the overall safety factor. This is because the cohesion and the internal friction angle have different reduction factors, and their effects on the overall safety factor are difficult to evaluate quantitatively.

While ISAKOV proposed the strength reduction shortest path method, he also defined the overall safety factor (as shown in formula (3)). We adopted this overall safety factor defined by ISAKOV in this paper. By substituting formula (2) into formula (3), the final expression of overall safety factor can be obtained, as shown in formula (4).

$$F_s = \frac{1}{1 - L\Big/\sqrt{2}} \tag{3}$$

$$F_s = \frac{1}{1 - \sqrt{\left(1 - \frac{1}{F_C}\right)^2 + \left(1 - \frac{1}{F_\varphi}\right)^2}\Big/\sqrt{2}} \tag{4}$$

3 3D Model Establishment and Result Analysis

In this paper, ABAQUS was used to calculate the model. To reduce the influence of the different buried depths of the tunnel vault, we set the model upper surface to be flat. The size of the model is shown in Fig. 2. The height, the span, the buried depth of the tunnel profile is 8.18 m, 10.79 m, and 10.79 m separately. Parameters of surrounding rock in the model: bulk density $\gamma = 20\text{kN/m}^3$, initial cohesion $c_0 = 150\text{kPa}$, initial internal friction angle $\varphi_0 = 24°$, elastic modulus $E = 1.3\text{GPa}$, Poisson's ratio $\upsilon = 0.35$.

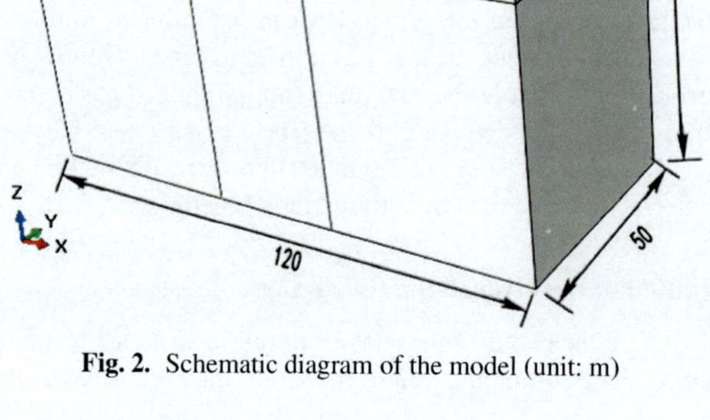

Fig. 2. Schematic diagram of the model (unit: m)

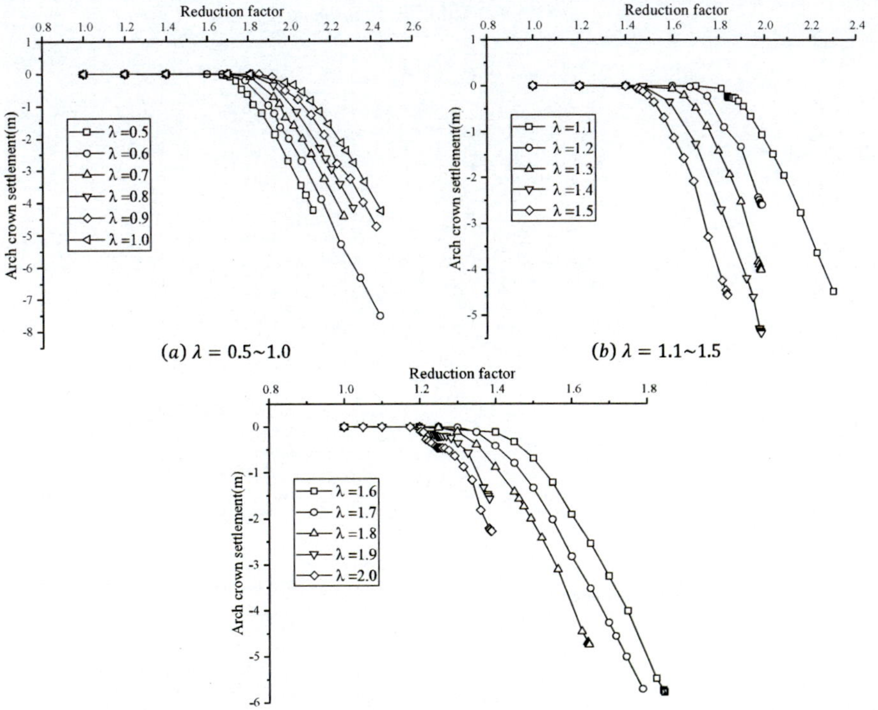

(*a*) $\lambda = 0.5\sim1.0$ (*b*) $\lambda = 1.1\sim1.5$ (*c*) $\lambda = 1.6\sim2.0$

Fig. 3. Arch crown settlement curves under different reduction ratios

The arch crown settlement at the tunnel portal section under different reduction ratios is shown in Fig. 3.

By substituting the reduction factors corresponding to displacement mutation points in Fig. 3 into formula (2) and (4), we got the lengths of reduction paths and the overall safety factors under different reduction ratios separately. The calculation results are shown in Table 1.

Table 1. The overall safety factors under different reduction ratios

λ	F_c	F_φ	L^2	F_s
0.5	1.741	3.481	0.689	2.421
0.6	1.788	2.979	0.635	2.292
0.7	1.813	2.589	0.578	2.162
0.8	1.860	2.325	0.539	2.079
0.9	1.918	2.131	0.511	2.022
1.0	1.981	1.981	0.491	1.981

(*continued*)

Table 1. (*continued*)

λ	F_c	F_φ	L^2	F_s
1.1	2.025	1.841	0.465	1.931
1.2	2.100	1.750	0.458	1.918
1.3	2.145	1.650	0.440	1.884
1.4	2.223	1.588	0.440	1.882
1.5	2.241	1.494	0.416	1.838
1.6	2.240	1.400	0.388	1.787
1.7	2.295	1.350	0.386	1.783
1.8	2.340	1.300	**0.381**	**1.775**
1.9	2.414	1.271	0.389	1.788
2.0	2.534	1.267	0.411	1.829

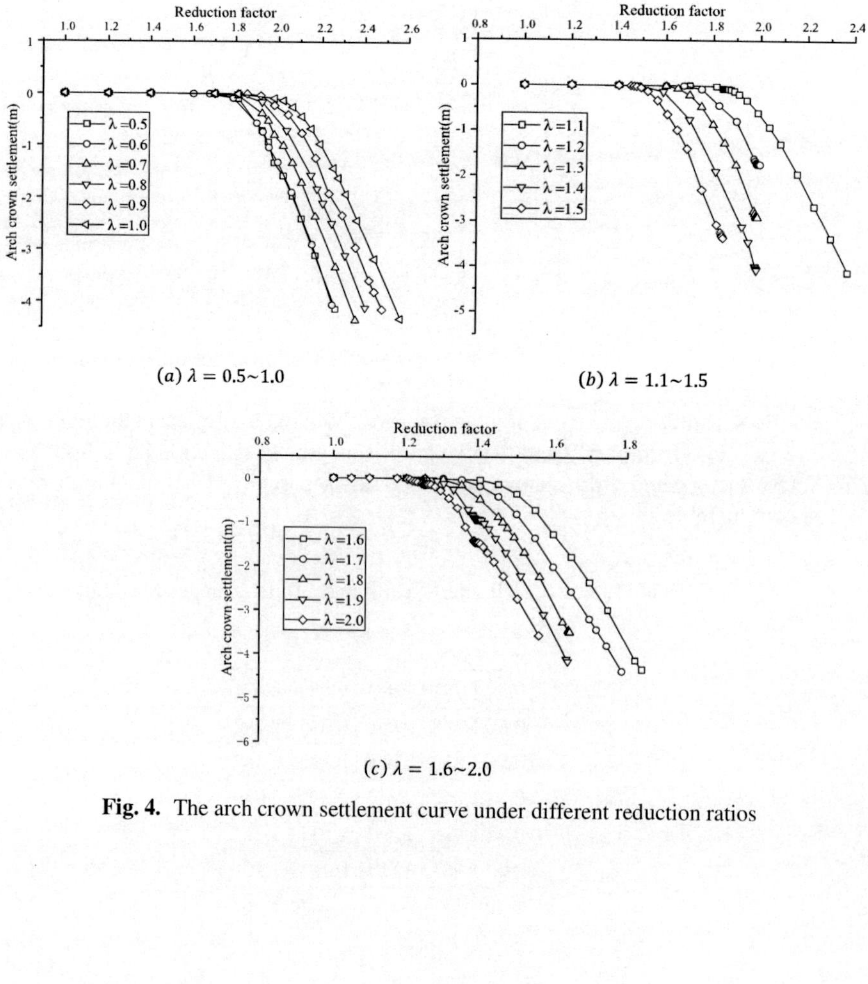

(*a*) $\lambda = 0.5\sim1.0$

(*b*) $\lambda = 1.1\sim1.5$

(*c*) $\lambda = 1.6\sim2.0$

Fig. 4. The arch crown settlement curve under different reduction ratios

It can be seen from Table 1 that as the reduction ratio λ increases, the reduction factor of cohesion gradually increases, while the reduction factor of internal friction angle gradually decreases. The reduction ratio λ and the overall safety factor F_s corresponding to the shortest path are 1.8 and 1.775, respectively. Research results show that the strength reduction shortest path method is suitable for the stability analysis of surrounding rock of the three-dimensional tunnel model.

The arch crown settlement of the tunnel excavation face ($Y = 50$) under different reduction ratios is shown in Fig. 4.

By substituting the reduction factors corresponding to displacement mutation points in Fig. 4 into formula (2) and (4), we got the lengths of reduction paths and the overall safety factors under different reduction ratios separately. The calculation results are shown in Table 2.

Table 2. The overall safety factors under different reduction ratios

λ	F_c	F_φ	L^2	F_s
0.5	1.813	3.625	0.725	2.514
0.6	1.900	3.167	0.693	2.430
0.7	1.925	2.750	0.636	2.293
0.8	2.038	2.547	0.628	2.275
0.9	2.045	2.272	0.574	2.155
1.0	2.108	2.108	0.552	2.108
1.1	2.137	1.943	0.518	2.037
1.2	2.190	1.825	0.500	1.999
1.3	2.275	1.750	0.498	1.996
1.4	2.310	1.650	0.477	1.954
1.5	2.346	1.564	0.459	1.920
1.6	2.400	1.500	0.451	1.905
1.7	2.465	1.450	0.450	1.901
1.8	2.520	1.400	**0.445**	**1.894**
1.9	2.600	1.369	0.451	1.905
2.0	2.676	1.338	0.456	1.914

It can be seen from Table 2 that the reduction ratio corresponding to the shortest path of the tunnel excavation face is 1.8, which is equal to that of the portal section. In both cases, the displacement mutation of the vault is taken as the instability criterion. However, the overall safety factor of the portal section is 1.894, which is greater than that of the excavation face (1.775). The above indicates that under normal circumstances, the portal section will lose stability first and is the most dangerous part of the tunnel.

In Fig. 5-a, the abscissa value represents the distance between the current section and the tunnel portal section ($Y = 0$), and the ordinate value represents the settlement of the vault. With the increase of the distance from the portal, the settlement of the tunnel vault decreases gradually. Figure 5-b and Fig. 5-c are the settlement diagram and the plastic deformation diagram separately. The tunnel portal section is located at the red box. The number of free faces of the tunnel portal is more than that of tunnel trunk and tunnel excavation face. According to the principle of material mechanics, the more free faces a material has, the more likely it is to break. It can be seen from the figures that the portal section has the largest vault displacement, and its plastic deformation of the arch waist is the largest. Therefore, it can be concluded that the portal section ($Y = 0$) is the dangerous section of the tunnel model.

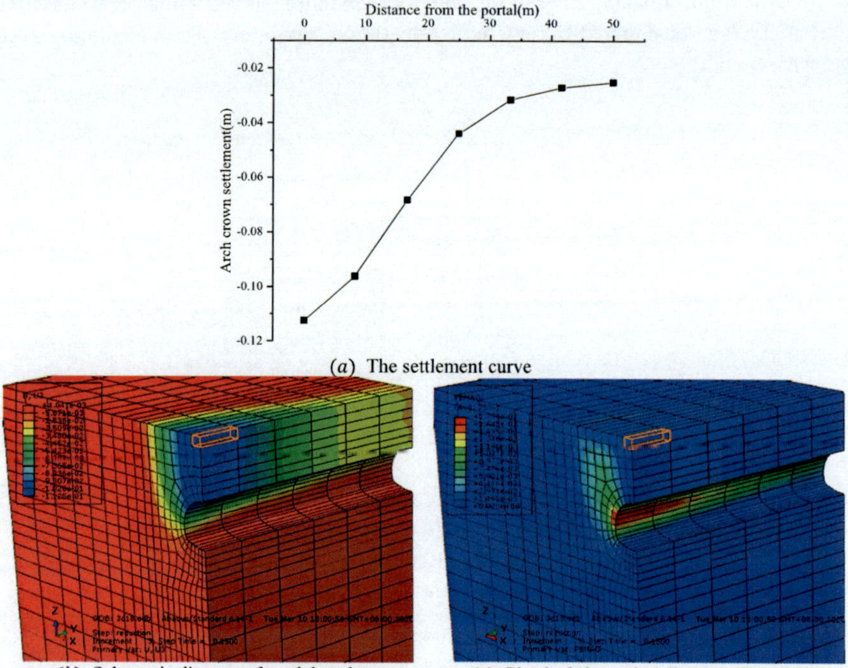

(*a*) The settlement curve

(*b*) Schematic diagram of model settlement

(*c*) Plastic deformation diagram of the model

Fig. 5. Arch crown settlement and plastic deformation at different distances from the portal section ($\lambda = 1.8$)

Figure 6 shows the distribution of surrounding rock deformation within the range of 0-50m from the tunnel portal. Because the constraint of the portal section is less than others in the tunnel, the displacement of surrounding rock in the portal is larger. And the displacement trend of vault and ground surface is subsidence. As the tunnel section moves from the portal to the tunnel excavation face, the surrounding rock displacement of the tunnel section shows a decreasing trend. The largest plastic deformation of all sections occurs at the arch waist. And the plastic deformation of the arch waist at the tunnel portal section is the greatest among all sections. With the tunnel section moving from the portal to the tunnel excavation face, the plastic deformation of the surrounding rock decreases. As the calculation time increases, the plastic deformation at the waist expands both in the X direction and the Y direction. However, it is faster to extend in the X direction than in the Y direction. Besides, the plastic zone in the two directions gradually extends to the ground surface.

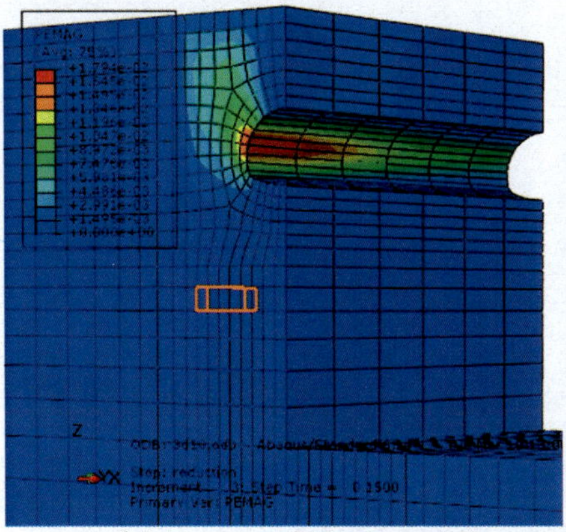

$(a)\ Time = 0.1500\ (F_{\varphi} = 1.3)$

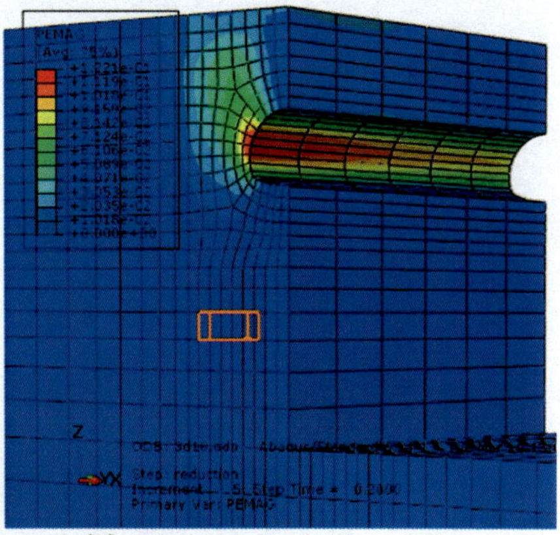

$(b)\ Time = 0.2000\ (F_{\varphi} = 1.4)$

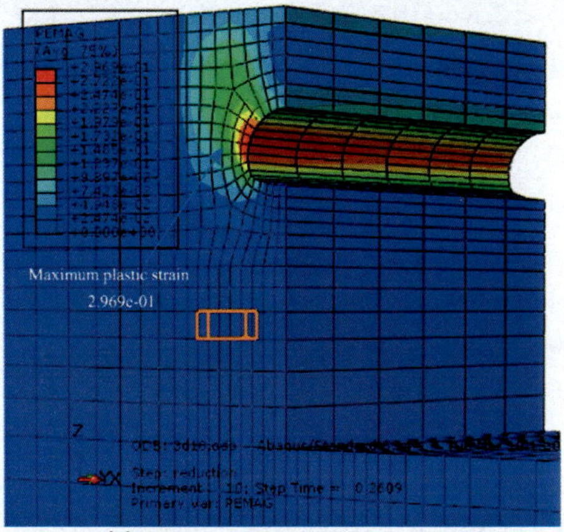

$(c)\ Time = 0.2609\ (F_{\varphi} = 1.52)$

Fig. 6. The plastic deformation diagram of the tunnel with the increase of calculation time ($\lambda = 1.8$)

The original equilibrium state of rock and soil mass at the portal is destroyed due to excavation. As a result, the rock and soil mass will have the corresponding displacement. When the vault displacement changes drastically, the tunnel vault and surrounding rock on the ground surface will collapse rapidly, which will lead to tunnel instability.

4 Influence of Strength Parameters on Stability of Surrounding Rock at the Tunnel Portal

4.1 Influence of Initial Internal Friction Angle

Four conditions of initial friction angle were set to be 20°,22°,24°, and 26°. The initial cohesion was set to be 150kPa, and the surrounding rock parameters were selected according to grade V surrounding rock. Based on the above information, the calculated data are shown in Table 3 and Table 4.

Table 3. Path lengths and overall safety factors at initial internal friction angles of 20° and 22°

$\varphi_0 = 20°$					$\varphi_0 = 22°$				
λ	F_c	F_φ	L^2	F_s	λ	F_c	F_φ	L^2	F_s
0.5	1.700	3.400	0.668	2.369	0.5	1.700	3.400	0.668	2.369
0.6	1.756	2.927	0.619	2.254	0.6	1.788	2.979	0.635	2.292
0.7	1.775	2.536	0.557	2.118	0.7	1.794	2.563	0.568	2.140
0.8	1.813	2.266	0.513	2.026	0.8	1.813	2.266	0.513	2.026
0.9	1.841	2.045	0.470	1.940	0.9	1.853	2.058	0.476	1.953
1	1.925	1.925	0.462	1.925	1	1.918	1.918	0.458	1.918
1.1	1.953	1.775	0.429	1.862	1.1	1.964	1.785	0.434	1.872
1.2	1.954	1.628	0.387	1.785	1.2	2.040	1.700	0.429	1.864
1.3	1.983	1.525	0.364	1.744	1.3	2.100	1.616	0.420	1.845
1.4	2.044	1.460	0.360	1.737	1.4	2.170	1.550	0.417	1.840
1.5	2.100	1.400	0.356	1.730	1.5	2.175	1.450	0.388	1.787
1.6	2.160	1.350	0.356	1.729	1.6	2.240	1.400	0.388	1.787
1.7	2.210	1.300	0.353	1.725	1.7	2.295	1.350	0.386	1.783
1.8	2.216	1.231	**0.336**	**1.695**	1.8	2.295	1.275	**0.365**	**1.746**
1.9	2.286	1.203	0.345	1.710	1.9	2.361	1.243	0.370	1.755
2	2.375	1.187	0.360	1.737	2	2.416	1.208	0.373	1.760

As can be seen from Table 3 and Table 4, the reduction ratio (corresponding to the shortest path of strength reduction) is the same for the four conditions with different initial internal friction angles, all of which are 1.8. The shortest path length (L_{min}) and the overall safety factor (F_s) corresponding to the four conditions all increase with the increase of initial internal friction angle (φ_0). When λ is 1.8, the vault settlement under four initial internal friction angle conditions is shown in Fig. 7. When the reduction factor is the same, the higher the initial internal friction angle is, the smaller the settlement of the arch crown is. When the displacement changes drastically, the reduction factor increases with the increase of the initial internal friction angle (φ_0).

Table 4. Path lengths and safety factors at initial internal friction angles of 24° and 26°

$\varphi_0 = 24°$					$\varphi_0 = 26°$				
λ	F_c	λ	F_c	λ	F_c	λ	F_c	λ	F_c
0.5	1.741	3.481	0.689	2.421	0.5	1.700	3.400	0.668	2.369
0.6	1.788	2.979	0.635	2.292	0.6	1.775	2.958	0.629	2.277
0.7	1.813	2.589	0.578	2.162	0.7	1.813	2.589	0.578	2.162
0.8	1.860	2.325	0.539	2.079	0.8	1.918	2.397	0.569	2.143
0.9	1.918	2.131	0.511	2.022	0.9	1.981	2.201	0.543	2.088
1	1.981	1.981	0.491	1.981	1	1.981	1.981	0.491	1.981
1.1	2.025	1.841	0.465	1.931	1.1	2.063	1.875	0.483	1.967
1.2	2.100	1.750	0.458	1.918	1.2	2.130	1.775	0.472	1.945
1.3	2.145	1.650	0.440	1.884	1.3	2.210	1.700	0.469	1.940
1.4	2.223	1.588	0.440	1.882	1.4	2.240	1.600	0.447	1.897
1.5	2.241	1.494	0.416	1.838	1.5	2.285	1.523	0.434	1.872
1.6	2.240	1.400	0.388	1.787	1.6	2.340	1.463	0.428	1.861
1.7	2.295	1.350	0.386	1.783	1.7	2.345	1.380	0.405	1.818
1.8	2.340	1.300	**0.381**	**1.775**	1.8	2.340	1.300	**0.381**	**1.775**
1.9	2.414	1.271	0.389	1.788	1.9	2.428	1.278	0.393	1.797
2	2.534	1.267	0.411	1.829	2	2.525	1.263	0.408	1.824

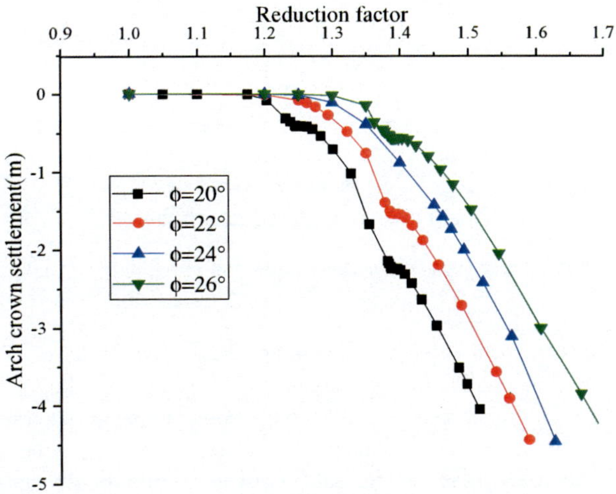

Fig. 7. Comparison of tunnel vault settlement under different initial internal friction angles (λ = 1.8)

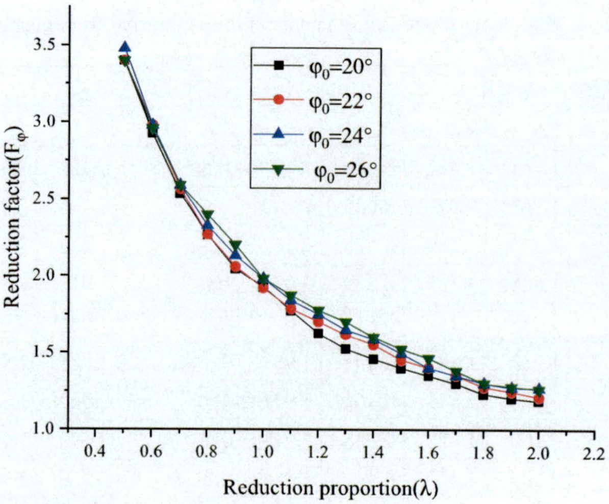

(*a*) The internal friction angle reduction factor F_φ

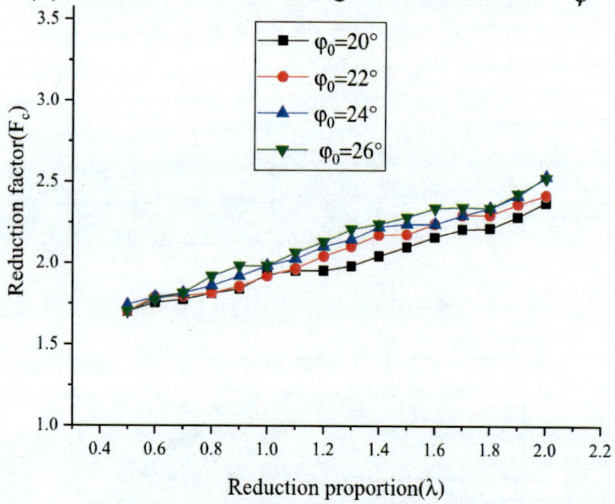

(*b*) The cohesion reduction factor F_c

Fig. 8. Reduction curves under different initial internal friction angles

It can be seen from Table 3, Table 4, and Fig. 8 that under the limit state, with the increase of initial internal friction angle, both the reduction factor of internal friction angle and the reduction factor of cohesion increase, but the growth rate of both gradually slows down. Therefore, in the actual project, when strengthening the surrounding rock, it is unreasonable to increase the internal friction angle blindly. Because when the internal friction angle exceeds a certain value, the improvement measures are neither economic nor ideal.

With the increase of the initial internal friction angle, although the reduction factor under the limit state increases, the growth rate of the reduction factor under different reduction proportions is not equal. The internal friction angle reduction factor in the range of $\lambda = 1.0 - 1.8$ in Fig. 8-a and cohesion reduction factor in the range of $\lambda = 1.2 - 2.0$ in Fig. 8-b are extracted, as shown in Fig. 9.

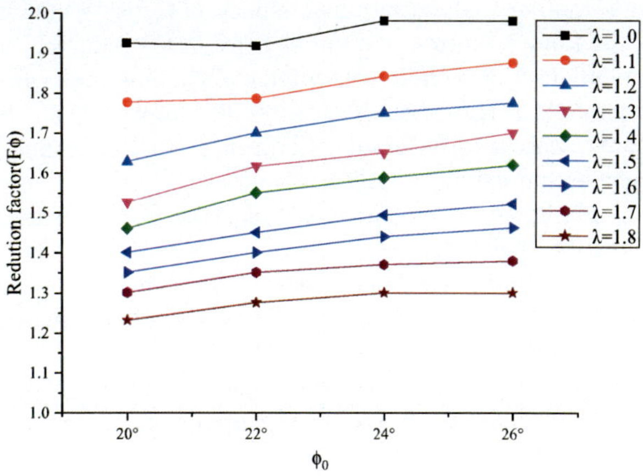

(a) The internal friction angle reduction factor F_{φ} $(\lambda = 1.0 - 1.8)$

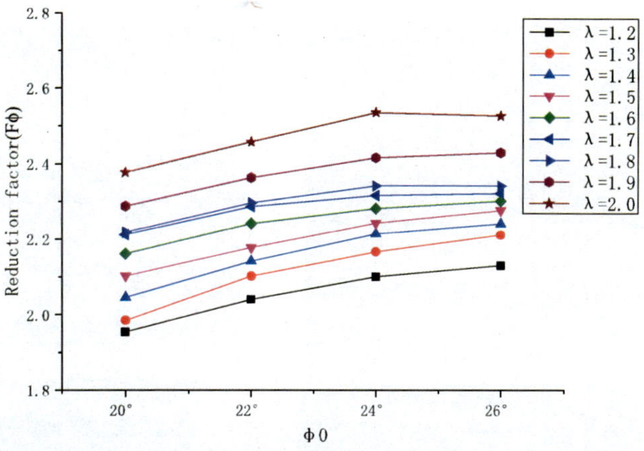

(b) The cohesion reduction factor F_c ($\lambda = 1.2 - 2.0$)

Fig. 9. Diagram of reduction factors under different reduction proportions

From Fig. 9-a, when the reduction ratio λ is in the range of 1.0–1.4, with the increase of the initial internal friction angle, the increase rate of the reduction factor F_{φ} is greater. That is, the internal friction angle in this range has a greater contribution to the resistance of the surrounding rock to deformation. From Fig. 9-b, when the reduction ratio λ is in the range of 1.4–1.6, with the increase of the initial internal friction angle, the increase rate of reduction coefficient F_c is greater. That is, the cohesion in this range contributes more to the resistance of the surrounding rock deformation.

The essence of the double strength reduction method can explain the above phenomenon. The essence of the double strength reduction method is to obtain the reduction factor of cohesion and the reduction factor of the internal friction angle separately (the two factors are not equal). Then the strength parameters of the rock and soil mass are reduced according to the determined reduction factors so that the rock and soil mass reaches the limit state from the initial state. In the reduction process, because of the different initial setting conditions (different reduction ratios), the coordination between strength parameters is complex. The analysis in this paper well reflects this essence. However, more research is needed on the influence of the reduction law of double strength parameters on the stability of surrounding rock.

Figure 10 shows the comparison of tunnel plastic deformation under different initial internal friction angles. It can be seen from the figure that with the increase of the initial internal friction angle, the failure form of the surrounding rock remains unchanged when the vault displacement changes drastically, and the plastic deformation shows a decreasing trend.

Figure 11 shows the comparison of tunnel settlement displacement under different initial internal friction angles. It can be seen from the figure that with the increase of the initial internal friction angle, the failure form of the surrounding rock remains unchanged when the vault displacement changes drastically, and the settlement shows a decreasing trend.

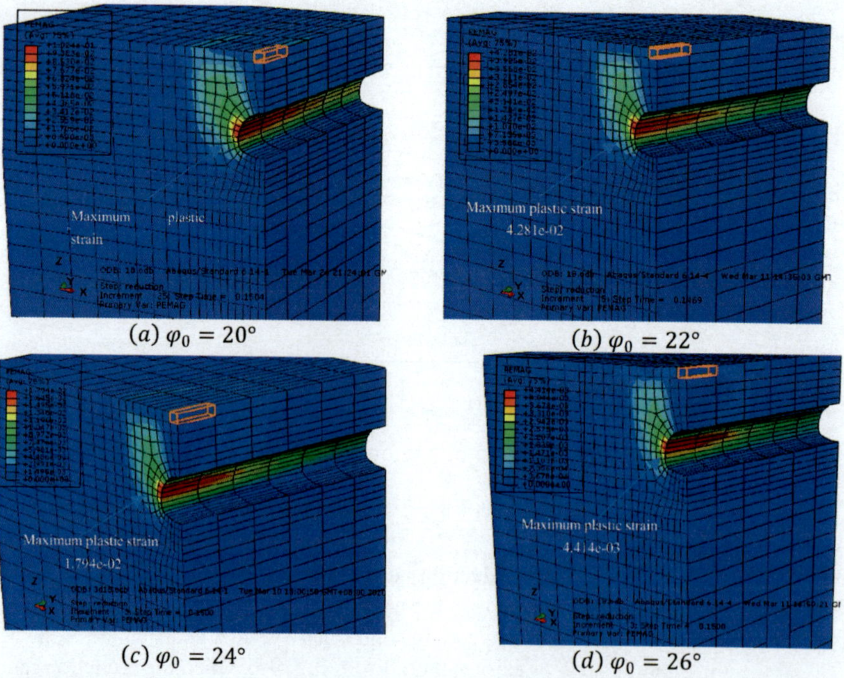

Fig. 10. The plastic deformation of the tunnel under different initial internal friction angles

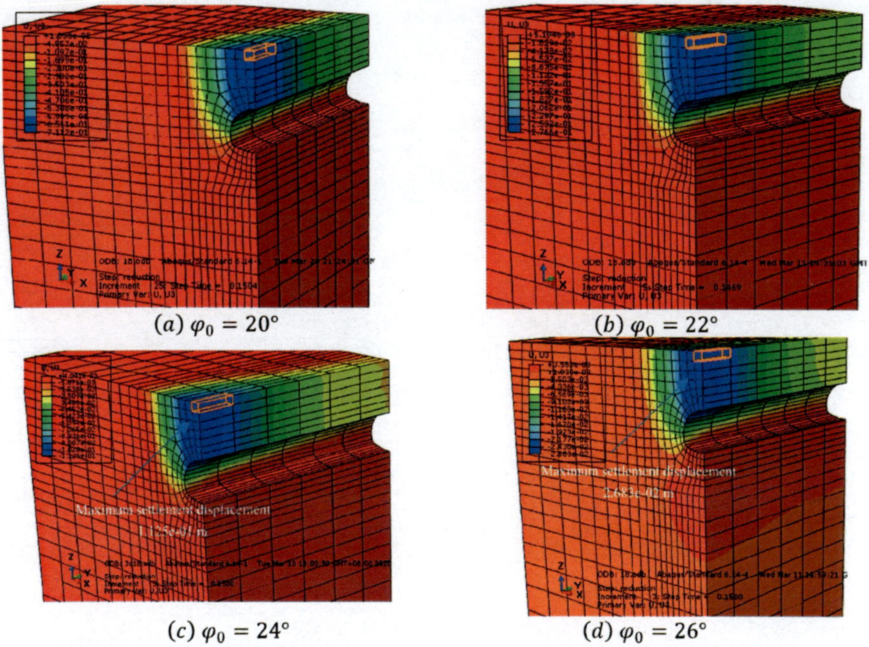

Fig. 11. Tunnel settlement displacement diagram under different initial internal friction angles

4.2 Influence of Initial Cohesion

Three conditions of initial cohesion were set to be 100 kPa, 150 kPa, and 200 kPa. The initial friction angle was set to be 24°, and the surrounding rock parameters were selected according to grade V surrounding rock. Based on the above information, the calculated data is shown in Table 5.

It can be seen from Table 3 that with the increase of initial cohesion c_0, the reduction ratio λ corresponding to the shortest path L_{min} increases. With the increase of initial cohesion c_0, both the shortest path length and safety factor show an increasing trend. The square of reduction path length L^2 and reduction ratio λ corresponding to different initial cohesion c_0 in Table 3 were extracted, and the extracted data were parabola fitted, as shown in Fig. 12.

In Fig. 12, the positions of the three parabolas gradually move up with the increase of the cohesion, which proves that the path length shows an increasing trend; The symmetry axis of the parabola also gradually moves to the right with the increase of the cohesion, that is, the reduction proportion corresponding to the shortest path increases.

It can be seen from Table 5 and Fig. 13 that under the limit state, with the increase of the initial cohesion, both the reduction factor of internal friction angle and the reduction factor of cohesion show an increasing trend. But the growth rate of the two will show a downward trend, which is consistent with the law under different initial internal friction angle conditions. Therefore, in the actual project, when strengthening the surrounding rock, it is unreasonable to increase the cohesion blindly. Because when the value of the cohesion exceeds a certain value, the improvement measures are neither economic nor ideal. With the increase of initial cohesion, although the reduction factor in the limit

Table 5. Path lengths and safety factors corresponding to different initial cohesions

	$c_0 = 100KPa$				$c_0 = 150KPa$				$c_0 = 200KPa$			
λ	F_c	F_φ	L^2	F_s	F_c	F_φ	L^2	F_s	F_c	F_φ	L^2	F_s
0.5	1.250	2.500	0.400	1.809	1.741	3.481	0.689	2.421	2.150	4.300	0.875	2.954
0.6	1.300	2.167	0.343	1.707	1.788	2.979	0.635	2.292	2.263	3.771	0.851	2.877
0.7	1.400	2.000	0.332	1.687	1.813	2.589	0.578	2.162	2.319	3.313	0.811	2.753
0.8	1.450	1.813	0.297	1.627	1.860	2.325	0.539	2.079	2.361	2.951	0.769	2.633
0.9	1.438	1.597	0.232	1.517	1.918	2.131	0.511	2.022	2.488	2.764	0.765	2.621
1	1.475	1.475	0.207	1.475	1.981	1.981	0.491	1.981	2.572	2.572	0.747	2.572
1.1	1.54	1.4	0.205	1.470	2.025	1.841	0.465	1.931	2.613	2.375	0.716	2.490
1.2	1.596	1.33	0.201	1.464	2.100	1.750	0.458	1.918	2.580	2.150	0.661	2.353
1.3	1.641	1.263	0.196	1.456	2.145	1.650	0.440	1.884	2.576	1.981	0.620	2.255
1.4	1.678	1.198	**0.191**	**1.447**	2.223	1.588	0.440	1.882	2.695	1.925	0.626	2.271
1.5	1.752	1.168	0.205	1.471	2.241	1.494	0.416	1.838	2.719	1.813	0.601	2.212
1.6	1.846	1.154	0.228	1.510	2.240	1.400	0.388	1.787	2.780	1.738	0.590	2.189
1.7	1.96	1.153	0.258	1.560	2.295	1.350	0.386	1.783	2.890	1.700	0.597	2.205
1.8	2.064	1.147	0.282	1.602	2.340	1.300	**0.381**	**1.775**	2.970	1.650	0.595	2.200
1.9	2.159	1.136	0.303	1.637	2.414	1.271	0.389	1.788	2.898	1.525	**0.547**	**2.097**
2	2.27	1.135	0.327	1.679	2.534	1.267	0.411	1.829	2.988	1.494	0.552	2.107

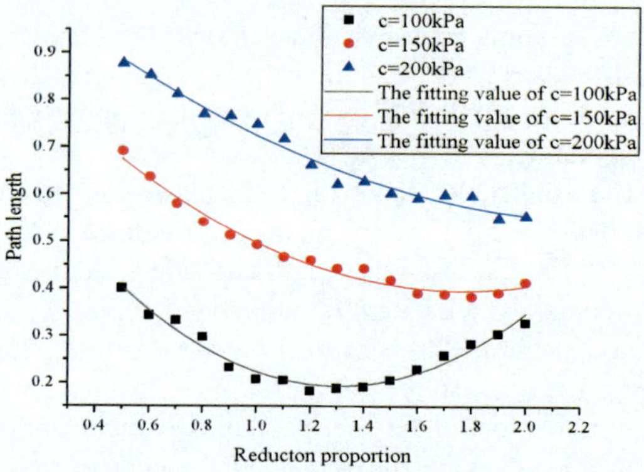

Fig. 12. Path lengths at different reduction proportions

state increases, the growth rate of the reduction factor in different reduction proportion ranges is not equal. Extract the reduction factor data of Fig. 13, as shown in Fig. 14.

From Fig. 14-a, when the reduction ratio λ is in the range of 0.5–1.1, with the increase of the initial cohesion, the increase rate of the reduction factor F_φ is greater. That is, the initial cohesion in this range has a greater contribution to the resistance of surrounding rock to deformation. From Fig. 14-b, when the reduction ratio λ is in the range of 0.5–2.0, with the increase of the initial cohesion, the growth rate of the reduction factor F_c

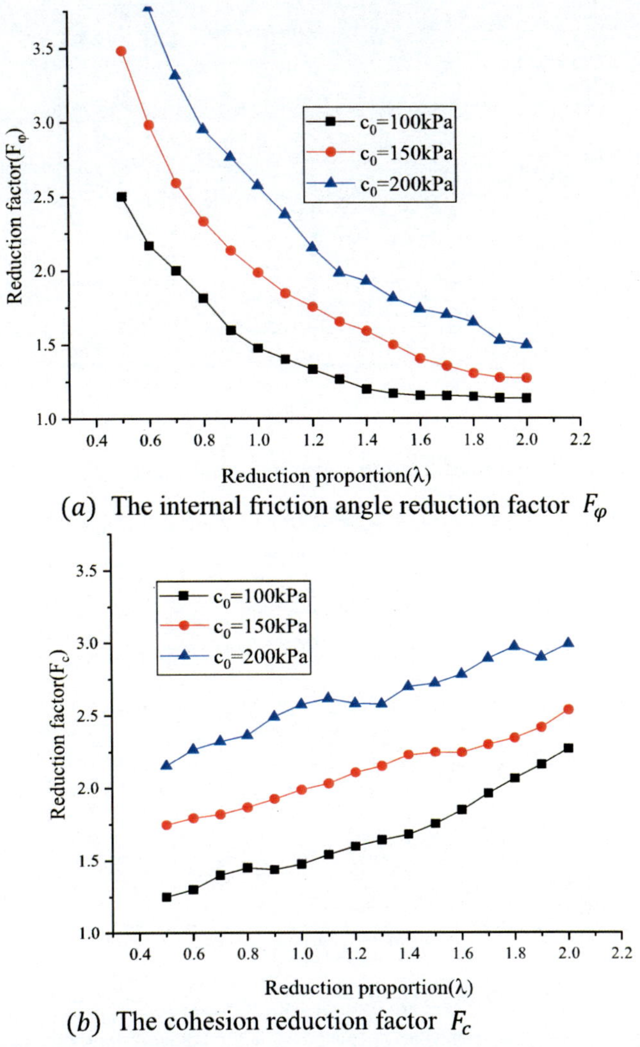

Fig. 13. Reduction curve under different initial cohesions

is nearly unchanged. That is, the reduction ratio λ will not have a significant impact on the increase of the reduction factor of cohesion.

Figures 15 and 16 show the tunnel plastic deformation and tunnel settlement displacement under different initial cohesions respectively. It can be seen from the figures that with the increase of the initial cohesion, the failure form of the tunnel surrounding rock remains unchanged when the vault displacement changes drastically, and the settlement shows a decreasing trend.

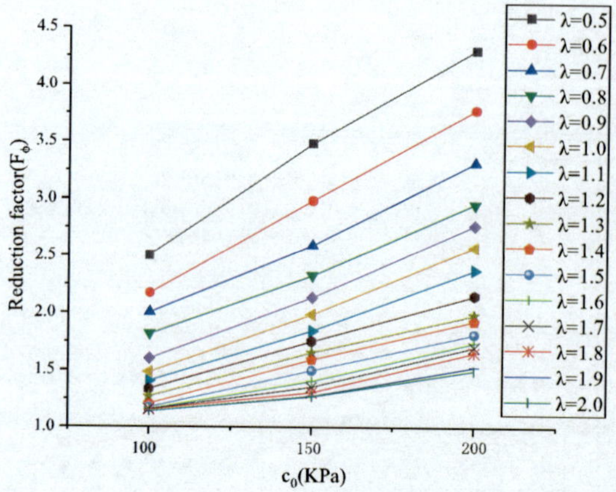

(a) The internal friction angle reduction factor F_φ $(\lambda = 1.0 - 1.8)$

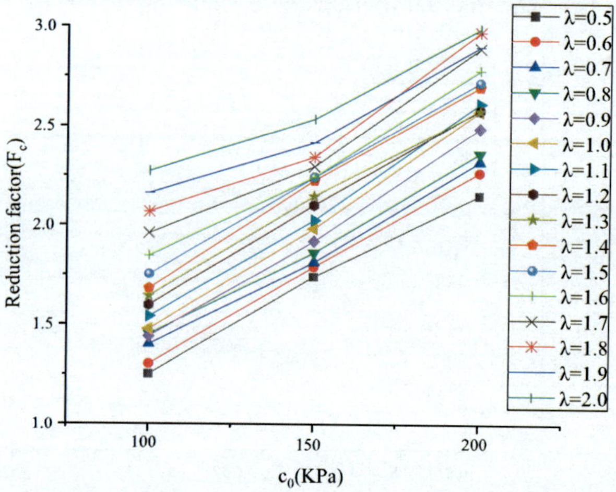

(b) The cohesion reduction factor F_c $(\lambda = 1.2 - 2.0)$

Fig. 14. Variation chart of reduction factors under different reduction ratios

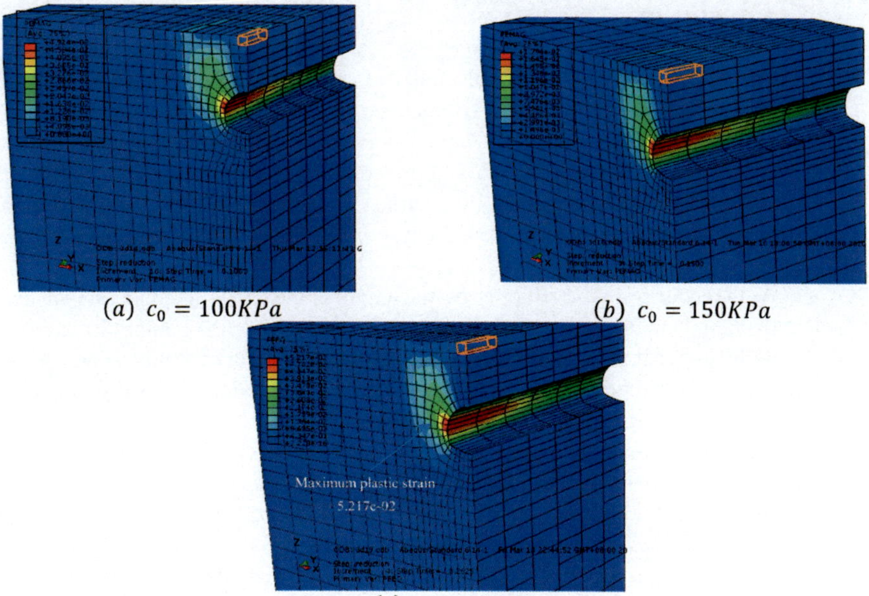

(a) $c_0 = 100KPa$ (b) $c_0 = 150KPa$

(c) $c_0 = 200KPa$

Fig. 15. Tunnel plastic deformation under different initial cohesions

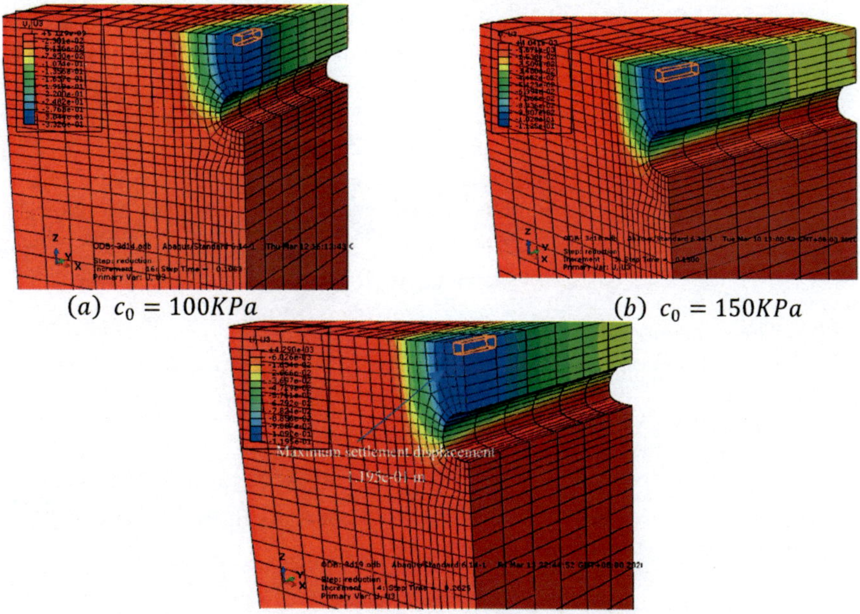

(a) $c_0 = 100KPa$ (b) $c_0 = 150KPa$

(c) $c_0 = 200KPa$

Fig. 16. Tunnel settlement displacement diagram under different initial cohesions

5 Conclusion

In this paper, a three-dimensional model of the tunnel portal section was established and analyzed. The stability analysis results of the surrounding rock at the tunnel portal verified the applicability of the strength reduction shortest path method to the three-dimensional model. Besides, the stability of tunnel surrounding rock with different strength parameters was studied by a numerical method, and the changing rules of the surrounding rock failure form were analyzed. The main conclusions are as follows.

1) As the reduction ratio λ increases, the length of the strength reduction path decreases first and then increases. And the variation conforms to the parabolic characteristics, indicating that there is the shortest path for the strength reduction. Thence the applicability of the strength reduction shortest path method in the finite element analysis of the three-dimensional tunnel model is verified.
2) It is reasonable to select the tunnel portal section as the most dangerous section in the three-dimensional model with a flat ground surface.
3) By comparing and analyzing the influence of strength parameters on the reduction factors, it was found that under the limit state, with the increase of strength parameters, both the reduction factor of internal friction angle and the reduction factor of cohesion increase. But their growth rate both shows a downward trend. Therefore, in the actual project, when strengthening the surrounding rock, it is unreasonable to increase strength parameters blindly. Because when the strength parameters exceed a certain value, the improvement measures are neither economic nor ideal.
4) Through the comparative analysis of the influence of strength parameters on the plastic deformation and settlement of the tunnel portal section, we found that the stability of the tunnel portal section increases with the increase of the strength parameters. But the change of the strength parameters has no significant effect on the failure mode of the tunnel portal section. In practical engineering, the reinforcement of the surrounding rock of the tunnel portal section can enhance the overall stability of the tunnel. But reinforcement measures can't change the distribution of dangerous parts. Therefore, continuous accident prevention should be carried out at the tunnel portal section.
5) The simulation in this study is based on the theoretical model and can predict the trend of soil deformation in case of tunnel failure. However, there may be some deviations from the actual tunnel conditions, and actual engineering simulations will be the next direction of research.

Acknowledgements. Project (2018JJ2519) supported by Hunan Provincial Natural Science Foundation of China.

References

1. Perazzelli, P., Leone, T., Anagnostou, G.: Tunnel face stability under seepage flow conditions. Tunn. Undergr. Space Technol. **43**, 459–469 (2014)

2. Anagnostou, G., Perazzelli, P.: Analysis method and design charts for bolt reinforcement of the tunnel face in cohesive-frictional soils. Tunn. Undergr. Space Technol. **47**, 162–181 (2015)
3. Lee, I.M., Lee, J.S., Nam, S.W.: Effect of seepage force on tunnel face stability reinforced with multi-step pipe grouting. Tunn. Undergr. Space Technol. **19**(6), 551–565 (2004)
4. Pan, Q., Dias, D.: Upper-bound analysis on the face stability of a non-circular tunnel. Tunn. Undergr. Space Technol. **62**, 96–102 (2017)
5. Griffiths, D.V., Lane, P.A.: Slope stability analysis by finite elements. Geotechnique **49**(3), 387–403 (1999)
6. Lian, Z., Han, G., Kong, X.: Stability analysis of excavation by strength reduction FEM. Chinese J. Geot. Eng. **23**(4), 407–411 (2001)
7. Zhang, P., Chen, Z.: Influences of soil elastic modulus and Poisson's ratio on slope stability. Rock Soil Mech. **27**(2), 299–303 (2006)
8. Tang, F., Zheng, Y.: Mechanism analysis on dual reduction factors about the progressive failure of slope. Chin. J. Undergr. Space Eng. **4**(3), 436–441+464 (2008)
9. Tang, F., Zheng, Y.: Analysis on safety reserve of slope with two strength reduction factor. J. Chongqing Jiaotong Univ. (Nat. Sci.) **26**(4), 95–100 (2007)
10. Tang, F., Zheng, Y.: Mechanism analysis on dual reduction factors about the progressive failure of slope. Chin. J. Undergr. Space Eng. **4**(3), 436–441+464 (2008)
11. Tang, F., Zheng, Y.: Effect on safety factors in different definitions based on strength margin. J. Civil Arch. Environ. Eng. **31**(3), 61–65+97 (2009)
12. Zheng, Y., Zhao, S., Song, Y.: Advance of study on the strength reduction finite element method. J. Logist. Eng. Univ. (03), 1–6 (2005)
13. Zhang, K.: Three-dimensional slope stability and failure analysis using shear strength reduction finite element method (SSR-FEM). Dalian University of Technology (2011)
14. Isakov, A., Korneyev, D.A., Moryachkov, Y.: Two-parameter criterion of road bed stability. In: Proceedings of Engineering Geology, Soil Mechanics and Foundations. Novosibirsk, Russia: [s. n.], pp. 31–38 (2010)
15. Isakov, A., Moryachkov, Y.: Estimation of slope stability using two-parameter criterion of stability. Int. J. Geomech. **14**(3), 613–624 (2014)
16. Chen, Z.: On Pan's principles of soil and rock stability analysis. J. Tsinghua Univ. Sci. Technol. **38**(1), 3–6 (1998)

Lessons and Mitigation Measures Learned from One Deep Excavation Failure Case

Wei Xiang, Yu-shan Luo(✉), and Zhi-rong Liang

Shanghai ShenYuan Geotechnical Co. Ltd., Shanghai 200011, China

Abstract. Deep excavations, most of those are temporary retaining structures, have not attracted enough attention from all parties involved in projects. Nowadays there have been numerous failures of deep excavations, and mostly have significant impacts on the safety of the excavation and the surrounding environment, especially in congested urban areas with poor hydrogeology and geology conditions. However, many failure cases remained unpublished and poorly explored due to complex reasons. The reported failure cases were too scarce to provide enough lessons to improve corresponding guidelines. Therefore the database needs more supplement cases. This paper briefly presents the failed case of a deep excavation in soft soil area in downtown city of Shanghai, China, which had an excavation area of about 25,000 m^2 and two-story basements. According to the field investigation, forensic studies and preliminary analysis, the main reasons for the failure was pointed out. Furthermore the re-checking calculations and mitigation measures were described briefly. The direct cost of repairing and rebuilding the failed excavation was more than 20 million RMB (about 2.8 million dollars), much more expensive than savings. Preserving some redundancy in design and construction is essential. This failed case report is expected to help deep excavations attract more attention and avoid similar failures.

Keywords: Deep excavations · Failures · Case histories · Forensic study · Mitigation measures

1 Introduction

With the rapid development of underground-space such as subway, tunnel, underground garage, underground commercial mall and so on, a number of excavations have emerged, and become much larger and bigger than before due to the scarce available ground resources in congested urban areas. However, deep excavation always has the potential to cause unfavorable effects on nearby ground as well as structures and facilities around it, especially in soft soils (Hu et al. 2003; Wang et al. 2010).

Being a temporary structure in most projects, the investment/cost for excavations and corresponding retaining structures is usually extremely restrained, and in consequence the reserved safety factors are much smaller than that of permanent structures. Although design and construction of excavations are stringently regulated by national and local codes, numbers of excavation failures (such as surrounding subsidence, groundwater

© The Author(s), under exclusive license to Springer Nature Switzerland AG 2021
S. Shu et al. (Eds.): GeoChina 2021, SUCI, pp. 92–104, 2021.
https://doi.org/10.1007/978-3-030-80142-7_7

inrushing, huge displacement of retaining walls and even fracture, etc.) still occur from time to time (Gong, *et al.* 2012; Jebelli et al. 2010). Therefore, it is necessary to summarize the experiences and lessons of the failures and provide reference to help avoid similar tragedies.

As field performance is a collective reflection of various factors involved in a real excavation, experience from the field performance of previous deep excavations provides a useful guidance in practical design. In recent decades, a series of studies helped us improve the understanding of the performance of deep excavations, and provided lots of experience when estimate the magnitude of movements or to check the rationality of the numerical analysis/field data of deep excavations. (Long 2001; Yoo 2001; Moormann 2004; Wang et al. 2010). However in the excavation database, failure reports are relatively limited, especially lacking the failure data in China. In spite of the fact that failures could provide a wealth of information to help improve excavation research, design, construction and management, many failure cases still remain unpublished and poorly explored due to problems of disputes and litigation. Unfortunately, the scarce reported failures failed to attract sufficient attentions of all parties involved in projects, failed excavations are still happening.

This paper briefly presents the failed case of a deep excavation in soft soil area in downtown city of Shanghai, China. According to the field investigation, forensic studies and preliminary analysis, the main reasons for the failure was pointed out. Furthermore the re-checking calculations and mitigation measures were described briefly. This failed case study shows how relevant factors "eat up" the safety factor and led to failure, so as to extend failure cases database and help deep excavations attract more attention and avoid similar failure cases.

2 Project Overview

The Failed case of the deep excavation was located in soft soil area in downtown city of Shanghai, China. The project comprised a two-story basements, and the excavation area and perimeter of the project were about 25000 m^2 and 480 m, respectively. The excavation depth of the site was mostly 8.7 m, where the site elevation of ground surface was leveled to not higher than 3.40 m in Wusong elevation (the elevation above Wusong Sea level). Unless otherwise specified, the elevation below is Wusong elevation.

Figure 1 shows the layout and surrounding environment of the failed project. The environment was not quite complex, surrounded by a small river (about 6 m–10 m in width and 0.4 m–0.9 m in depth) to the north, a 28 m-wide municipal road to the east, a public green land to the west and internal construction site (the construction work of the south 1-story basement was finished, up to ground level) to the south. The main protection objects around the project were the municipal road and underground pipelines in the East.

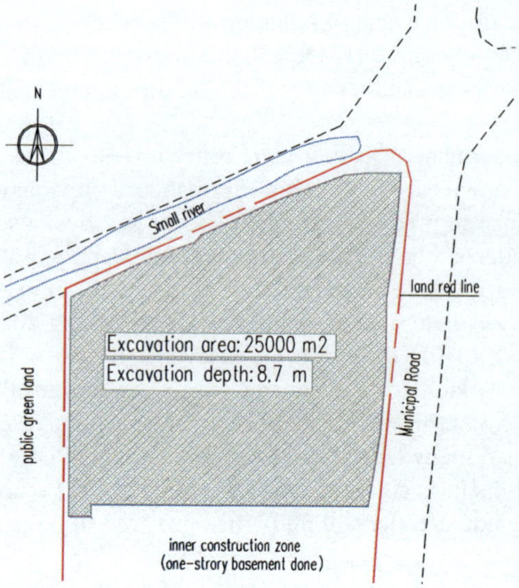

Fig. 1. Sketch of the surroundings around excavations

2.1 Geology and Ground Condition

The project site was situated at Yangtze River Delta alluvial plain. According to the geotechnical investigation report and the division work of Gao et al. (1986) the subsurface soils at the construction site were mainly thick soft soils comprising Quaternary alluvial and marine deposits. As shown in Table 1, from ground surface to a depth of about 30 m, the soil profile could be divided into 4 layers, among which Layer ② and Layer ⑤ could be subdivided into 2 sub-horizontal layers, respectively. The top layer is loose fill mixed with lots of construction waste, having relatively poor engineering mechanical properties. The Layer ②1 and Layer ②3 are over-consolidated and commonly referred to as stiff surface crust. In this site, Layer ③ was lacked. Layer ④, where the excavation was located on, was soft clay layer of thickness about 8.5 m–11.0 m. The soft clay has relatively higher natural water content, void ratio and compressibility but lower shear strength, which is one of the main unfavorable factors for deformation control for excavations in soft areas. Underlying Layer ④ was Layer ⑤1-1 and Layer ⑤1-2. The fifth layer was mainly gray silty clay a with medium plastic and medium to high compressibility. The physical and mechanical properties of Layer ⑤ were much better than that of Layer ④.

2.2 Original Excavation Design Scheme

In the original excavation design, the excavation was constructed using the bottom-up construction method, bored piles and waterproof curtain were designed to be the retaining system. Meanwhile, the excavation was supported by one level of reinforced concrete strut. The section size of main RC struts was 1000 mm in width and 800 mm in height.

Table 1. Physico-mechanical parameters of the soils

soil sequence and description	h (m)	γ (kN/m³)	φ (°)	C (kPa)	Kv (cm/s)	Kh (cm/s)	k* (cm/s)	q_c(MPa)
①Miscellaneous Fill	1.8	18.0	10.0	10.0	/	/	5.0E−4	0.65
②1 Silty clay	1.8	18.4	19.5	21.0	2.73E−7	8.63E−7	9.0E−6	0.74
②3 Sandy silt	3.3	18.6	31.0	5.0	6.94E−5	1.39E−4	4.1E−4	1.81
④ Soft clay	10.4	16.8	11.5	11.0	8.03E−8	8.95E−8	6.5E−7	0.52
⑤1–1 Silty clay	4.7	18.1	18.5	15.0	3.10E−7	1.94E−6	6.0E−6	0.71
⑤1–2 Silty clay	4.3	18.2	19.0	16.0	4.83E−7	2.10E−6	8.5E−6	0.98

Note: Underlined data mean empirical values, while the others mean standard values. h = thickness of typical section; γ = unit weight; φ = angle of internal friction obtained from direct shear test; c = cohesion obtained from direct shear test; kv = vertical coefficient of permeability; kh = horizontal coefficient of permeability; k* = recommended coefficient of permeability. $\mathbf{q_c}$ = measured cone penetration test end resistance.

And the horizontal spacing of the struts was approximately 10 m. Figure 2 presents a typical section of retaining structures.

The bored piles was designed embedding into the fifth layer to help retaining soil, with 900 mm in diameter and 18.3 m in length. Moreover the bored piles were filled with concrete strength grade of 30 (the underwater casting need to be improved by one level). Herein the concrete strength ($f_{cu,k}$) refers to the compressive strength of molded 150 mm cubes after 28 days' curing should be more than 30 MPa, according to Chinese code MCPRC (GB50010–2010, 2015). This concrete grade C30 is somewhat equivalent to $f_{ck,cub}$ = C25/30 according to European Standard CEN (BS EN 1992-1-1, 2000).

The groundwater level in Shanghai is generally located at 0.5–1.0 m below the ground surface and the most unfavorable groundwater level (i.e. 0.5 m) was used in design. Since the confined aquifer is very deep in the subsoil compared with the excavation depth, the confined water pressure is not a risk in this project. Consequently two rows of deep cement-soil mixing wall mixed by dual-anger mixer (Ø700 mm @500 mm) were designed as waterproof curtain, to cutoff the phreatic water during excavation construction. The waterproof curtain was designed 15.6m in length and 13% in cement slurry by mass.

The lateral supporting RC strut was designed 2.1 m below ground surface, 6.6 m above the bottom. The horizontal spacing of the temporary struts was mostly 6 m–10 m, with concrete grade of C30.

Since the top 15 m soil deposits in this site are quite soft, the soils under the final cutting surface need some ground improvement to increase stiffness and help control deformations, especially in the middle of each side and those having deep water-collecting wells inside. The weak soils were designed mixing with cement mixed by dual-anger mixer (Ø700 mm @500 mm, 13% in cement slurry by mass).

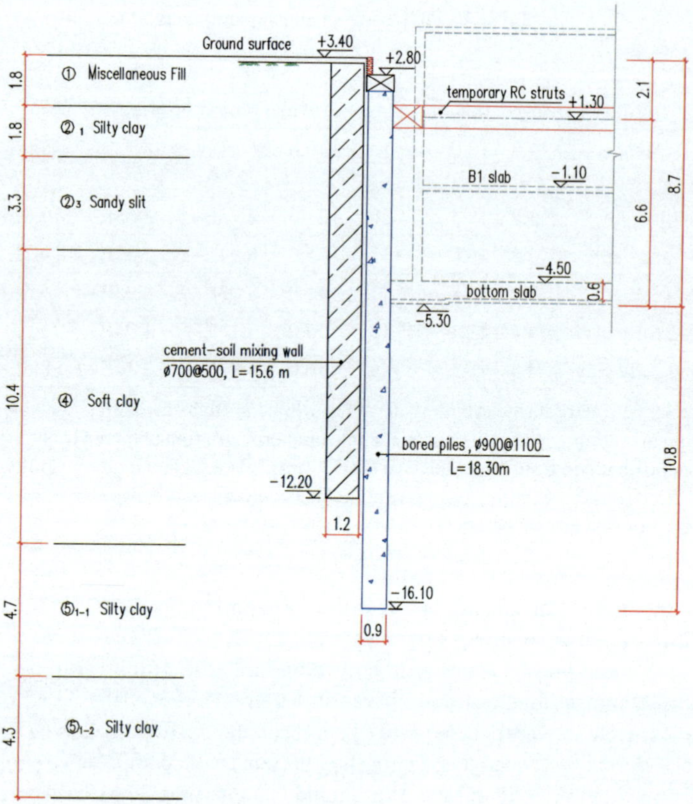

Fig. 2. Typical section of retaining structures at the failed zone

3 Excavation Failure Process and Forensic Studies

3.1 Excavation Failure Process

After the construction of central posts, deep cement-soil mixing waterproof curtain, bored piles, ground improvement, and lateral RC struts (curing for over 7 days), the excavation work started from northeastern part of the site. With the northeastern part excavated to bottom and construction of plain concrete cushion was just set out, southeastern excavation was carried out. When the southeastern part was just excavated to bottom and the northeastern excavated to edge of the retaining piles, large displacement of retaining bored piles accumulated shifting towards excavation, central posts were gradually uplifted, and three RC struts in the northeastern corner were suddenly broken (see in Fig. 3), followed by the municipal road nearby collapsed, with one of the water supply pipelines damaged seriously. The water from the damaged pipeline rushed into the excavation, fortunately, no human-casualties were reported. View of the broken RC struts and flooding excavation could be seen in Fig. 3.

Fig. 3. Picture of the failed deep excavation

In order to avoid further expansion of the settlement and damage, the emergency plan was adopted in-time. The northeastern and northwestern part was temporary backfilled until the deformation was stable. Specially, the areas closed to retaining piles were backfilled with gravel soil (adding some cement slurry as soil improvement) to help control deformation. The distribution range of backfill treatment could be seen in Fig. 4.

3.2 Field Investigation and Testing

The failed excavation had drawn lots of attentions from different parties involved in this excavation construction, not limited to the construction investor, designer, monitor, and constructor and so on. With a series of investigation and testing conducted by three different independent third parties (ranging from non-professional operations, soil parameters, design reviews, retaining system tests, monitoring data reviews), the reasons of the failed excavation were carefully studied and analyzed.

3.3 Forensic Studies and Discussions for the Excavation Failure

Here the "forensic study" is similar to reverse engineering, namely the end result is known and one then looks for plausible reasons for the failure. After preliminary analyses, the contributing reasons of the failed excavation were conclude as follows.

1) Insufficient length of retaining bored piles

Core drilling tests were carried out in 8 retaining bored piles, and the investigated length were compared with original design length, as shown in Table 2. The first 3 piles, mostly

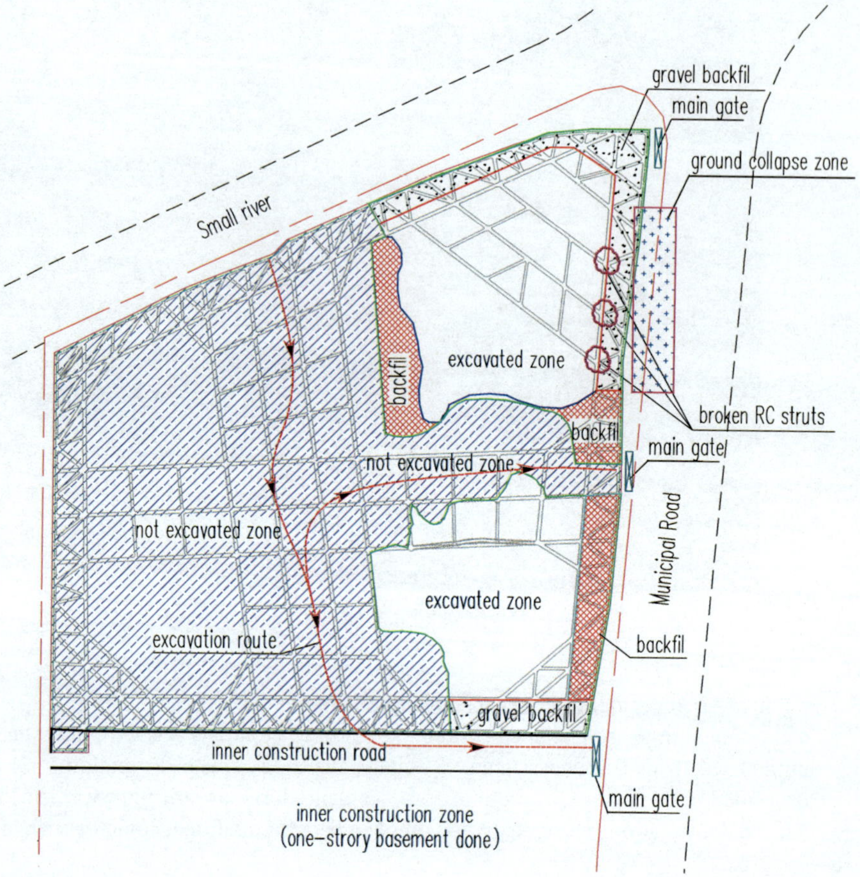

Fig. 4. Schematic diagram of the failed excavation (broken RC struts was marked by circles) and temporary backfill treatment

located in eastern part, were tested by Company A and the 4th-8th pile (located in north, west and south) were tested by Company B. Note that the A-1# and A-2# piles were located in the collapse area where the displacement of retaining piles were quite large, the coring operation was incomplete and discontinuous. Hence, the length of this two piles were estimated from the length of steel reinforcement cage.

Table 2 indicates that the length of tested retaining bored piles were all insufficient, 0.46 m–1.60 m shorter than original design. According to the geotechnical investigation report, the elevation of bottom of Layer ④ was -12.05 m--14.19 m. The shortest tested piles probably embedded only 0.31 m underneath Layer ⑤1-1, where the insufficient embedded depth was too small to provide anchoring force. Large deformation thus generated quickly from the bottom and finally led to “kicking” failure of the whole excavation.

Table 2. Comparison of tested and designed length for retaining bored piles

Test No	Elevation of pile-bottom (m)		Length of retaining pile (m)		Difference between tested and designed length (m)
	Designed	Tested	Designed	Tested	
A-1#	-16.1	-14.9	18.3	17.1	-1.2
A-2#	-16.1	-14.6	18.3	16.8	-1.5
A-3#	-16.1	-14.5	18.3	16.7	-1.6
B-4#	-16.1	-14.59	18.3	16.79	-1.51
B-5#	-16.1	-15.21	18.3	17.41	-0.89
B-6#	-16.1	-15.13	18.3	17.33	-0.97
B-7#	-16.1	-15.64	18.3	17.84	-0.46
B-8#	-15.5	-14.92	16.4	15.82	-0.58

Note: (1) The top elevation of retaining piles was + 2.20;
(2) Underlined data were inferred from the length of steel reinforcement cage;
(3) Difference between tested and designed pile length=length of tested pile – length of designed pile.

2) Inadequate strength of retaining bored piles

Table 3 lists the tested concrete compressive strength ($f_{cu,k}$) of the core sample from core drilling tests conducted by two different independent third party. The cylindrical strength (f_{ck}) is about 24% lower than cubic strength ($f_{cu,k}$) in Shanghai according to Chinese national code MCPRC (GB50010–2010, 2015). Therefore the tested compressive strength ($f_{cu,k}$) listed in Table 3 was calculated by tested cylindrical strength (f_{ck}) multiplied 0.76, i.e., $f_{cu,k} = 0.76f_{ck}$.

Note that the tested compressive strength values of the 1st and 2nd piles (locations near the collapse zone) were much lower than others, it is possible that the piles got damaged more or less, thus these two results were not accurate to some extent and recommended not to be included in the statistics.

As shown in Table 3, about 75% of the tested retaining bored piles had inadequate compressive strength, and the difference with original design C30 was about 1.3 MPa–6.3 MPa. Inadequate strength of piles will lead to lower Ec and further lower bearing capacity against horizontal earth pressure, which is another factor contributing to the failure.

3) Poor quality of reinforced concrete struts

The original design required the RC struts to be straight so as to decrease the external forces. However, the in-situ investigation found out that at least 21% of the struts in excavated zone had deflection like fish belly beyond the requirements of design and standards. In the northeast side of the excavation where struts had visible ruptures or even broken, a series of investigations detecting the reinforcement and concrete qualities were carried out in 15 different RC struts by Company C. The investigation indicated that the reinforcement arrangement and quantity of the tested struts basically meet the original

Table 3. Comparison of tested and designed strength ($f_{cu,k}$) for retaining bored piles

Test No	Concrete compressive strength $f_{cu,k}$ of piles (MPa)		Difference between tested and designed concrete compressive strength of piles (MPa)
	Designed	Tested	
A-1#	30.0	16.68	-13.32
A-2#	30.0	15.89	-14.11
A-3#	30.0	23.70	-6.30
B-4#	30.0	23.90	-6.10
B-5#	30.0	28.70	-1.30
B-6#	30.0	27.80	-2.20
B-7#	30.0	32.00	2.00
B-8#	30.0	31.00	1.00

Note: (1) Underlined data were probably disturbed by collapse, not included in statistics and forensic studies;
(2) Difference between tested and designed concrete compressive strength of piles= $f_{cu,k}$ of tested pile $-f_{cu,k}$ of designed pile.

design requirements. However, there were 5 RC struts (over 30% of the tested sample) having insufficient concrete strength, the difference with original design (concrete grade C30) was about 0.3 MPa–7.3 MPa. For the strut having $f_{cu,k}$ of 22.7 MPa, the maximum axial compressive force it could bear was only 75.7% of the original design.

In addition, quite a lot of RC struts (over 20%) had visible missing edges and corners. Smaller section area of RC struts will inevitably lead to less bearing capacity, even fail to meet the design requirements. Therefore, when there is a dynamic/impact load or the retaining bored piles had larger deformation, RC struts having low straightness would have larger eccentric loads and bending moment. Compounding the problems are the defects of struts (missing edges and corners, eccentric error, inadequate strength). Gradually the struts will be less effective, leading to a decrease in the stability of the retaining system, and even developed to failure.

4) Invalid monitoring data

Overall, the environment of this project is not complex and even kind of loose, only the municipal road and adjacent pipelines needed more protection. Whereas, good environment was directly related to the failure. Through recalculation and back analysis on data got from in-situ investigations, the maximum deflection of the retaining wall was over 50 mm, the central posts were uplifted over 30 mm, and the deformation rate exceeded 3 mm/d, which already surpassed the allowable value according to original design and the Shanghai code (SUCTC, DG/TJ08–61-2018, 2018). However, in the monitoring report, data were tampered to be normal. It was said that some measuring points were not actually monitored for several days, which means the monitoring data were somewhat invalid, and did not truly reflect the real performance of the excavation. Lack of vigilance and invalid monitoring data resulted in ignoring the unusual performance of the excavation, the rate and amount of deformation increased more and more, finally the accident occurred.

4 Re-checking Calculation and Mitigation Design

Before further measures were taken to repairing or even rebuilding the excavations, a series of re-checking calculations were conducted with the program, FRWS 8.2 (2018). For this project, the limit equilibrium method was used for overall stability of retaining system calculation, moreover, the horizontal coefficient of subgrade based on elastic foundation beam was chose to predict displacement based on the Chinese code and Shanghai local standard (MCPRC, JGJ120–2012,2012 and SUCTC, DG/TJ08–61-2018, 2018).

For this excavation, the required factors of stability are as follows:

- $F_{so} = 1.3, F_{sb} = 1.6$, Safety Level II, according to Chinese standard (MCPRC, JGJ120–2012,2012);
- $F_{so} = 1.25, F_{sb} = 1.9$, Safety Level II, according to Shanghai local standard (SUCTC, DG/TJ08–61-2018, 2018).

Typical results of re-checking calculations were shown in Table 4 and 5.

Table 4. Parameters for different models in re-checking calculation

Model No	Length of retaining bored pile (L/m)	Strength of retaining bored pile ($f_{cu,k}$/Mpa)	Stiffness of reinforced concrete struts (EA/kN)	Surcharge on the municipal road (P/kPa)	Remarks
M1	18.3 m	30	2.16×10^7	20	Original design
M2	**16.7 m**	**20**	**1.12×10^7**	**30**	Worst condition (collapsed zone)
M3	16.7m	20	2.16×10^7	10	Mitigation measure A
M4	20.3 m	30	**1.12×10^7**	10	Mitigation measure B

Table 5. Calculated results of different re-checking models

Model No	Maximum wall horizontal displacement (δ_{hm}/ mm)	Maximum ground settlement (δ_{vm}/ mm)	Maximum wall bending moment (M/ kN.m/m)	Axial force of the struts (F/ kN/m)	Factor of overall stability (F_{so})	Factor of stability against basal heave (F_{sb})
M1	31.9	39.5	1259.7	351.5	1.93	1.59
M2	38.8	51.2	1239.1	387.7	1.57	1.27
M3	23	31.1	745.9	183.5 (1st)	1.75	1.39
				373.1 (2nd)		
M4	21.1	25.7	794.3	198 (1st)	2.45	1.96
				356.8 (2nd)		

Note: (1) Underlined data are the axial force of the second struts (added);
(2) In M2~M4, the parameters of miscellaneous fill (mainly φ and c) were discounted to some extent, taking the effect of collapse into consideration.
(3) F_{so}= ratio of anti-sliding moment to sliding moment based on Swedish slice method; F_{sb}= ratio of the resistant shear force to the driving shear force along a certain slip surface, based on Prandtl theory.

After several trials, the mitigation measures for this failed excavations (mainly the typical section as shown in Fig. 2) could be divided into categories: Mitigation measure A and B.

- **Mitigation measure A** (for zones bored piles were still effective):
  1) repair the visible cracks, missing edges and corners of the RC struts; cut off and rebuild the broken struts; 2) add another level of RC struts; 3) restrict the surcharge on surrounding road to no more than 10 kPa; 4) add trestles, platforms and corresponding central posts to help organize construction works inside the excavation.
- **Mitigation measure B** (for zones where bored piles were broken or not effective):

Besides the mitigation measures in Mitigation measure A, the following measures are still needed (refer to Fig. 5).

1) clear underground obstacles and increase ground stiffness with deep cement-soil mixing wall; 2) install new bored piles inside the deep cement-soil mixing wall; 3) improve soils between the original failed retaining pile and new retaining piles, and those inside excavation (under the bottom).

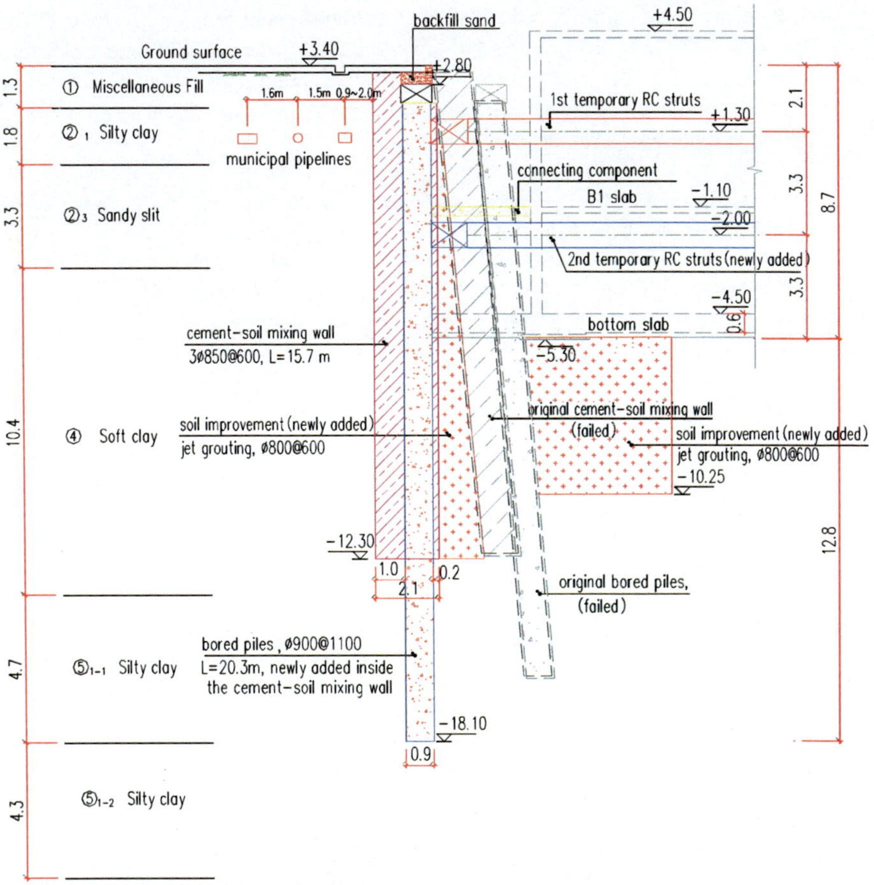

Fig. 5. Typical mitigation measures of retaining structures at the failed zone (Mitigation measure B)

5 Concluding Remarks

This paper presents a failed case of a deep excavation in soft soil area in downtown city of Shanghai, China. After field investigation, forensic studies and preliminary analysis, four main reasons of excavation failure were concluded: (1) insufficient length of retaining bored piles led to poor anchoring and kicking failure; (2) inadequate strength of retaining bored piles; (3) poor quality of reinforced concrete struts in straightness, strength and section area; (4) invalid monitoring data and ignoring the unusual performance of the excavation.

Further consulting and mitigation design were carried out with a series of re-checking calculations based on limit equilibrium approach and elastic foundation beam method. Finally, the direct cost of repairing and rebuilding the retaining system was more than 20 million RMB (about 2.8 million dollars) in total, more expensive than savings of jerry works and original design optimization. Another important lessons could be learn from this case are summarized as follows. Although retaining systems for excavation are temporary, the optimization design, jerry-build construction and other relevant factors together "eat up" the safety factor, good design and construction should still preserve rational or relevant redundancy. The cost of this redundancy is far less than the cost of failure and increased repair or even rebuild.

Acknowledgements. The authors are indebted to the Science and Technology Committee of Shanghai for its financial supporting for this study (No. 20QB1404500). Also, we appreciate the kind help of some other companies in providing in-situ information and data. Comments from reviewers and editors to improve the clarity and quality of the paper are welcome and will be appreciated.

References

CEN (European Committee for Standardization). Concrete-Part1: Specification, performance, production and conformity. BS EN206–1 (European Standard English version). BSI (2000)

FRWS/2018. Standard user's manual-version 8.2[in Chinese].QiMSTAR (Tongji Qimingxing, Inc.), Shanghai, China (2018)

Gao, D.Z., Wei, D.D., Hu, Z.X.: Geotechnical properties of Shanghai soils and engineering applications. ASTM Special Technical Publication, pp. 161–177 (1986)

Gong, X.-N., Zhang, X.-C.: Excavation collapse of Hangzhou subway station in soft clay and numerical investigation based on orthogonal experiment method. **13**(10), 760–767 (2012)

Hu, Z.F., Yue, Z.Q., Zhou, J., et al.: Design and construction of a deep excavation in soft soils adjacent to the Shanghai metro tunnels. Can. Geotech. J./Rev. Can. Geotech. **40**(5), 933–948 (2003)

Jebelli, J., Meguid, M.A., Sedghinejad, M.K.: Excavation failure during micro-tunneling in fine sands: a case study. Tunnel. Undergr. Space Technol. **25**(SI), 811–818 (2010)

MCPRC (Ministry of Construction of the People's Republic of China). Code for design of concrete structures GB 50010-2010 (revised edition, 2015). China Architecture and Building Press, Beijing (2015). [in Chinese]

MCPRC (Ministry of Construction of the People's Republic of China). Technical specification for retaining and protection of building foundation excavations. JGJ120–2012. China Architecture and Building Press, Beijing (2012). [in Chinese]

Long, M.: Database for retaining wall and ground movements due to deep excavations. **127**(3), 203–224 (2001)

Moormann, C.: Analysis of wall and ground movements due to deep excavations in soft soil based on a new worldwide database. Soils Found. **44**(1), 87–98 (2004)

SUCTC (Shanghai Urban-rural Construction and Transportation Commission). Technical code for excavation engineering. DG/TJ08–61–2018. Tongji University Press, Shanghai (2010). [in Chinese]

Wang, J.H., Xu, Z.H., Wang, W.D.: Wall and ground movements due to deep excavations in Shanghai soft soils. J. Geotech. Geo-environ. Eng. **136**(7), 985–994 (2010)

Yoo, C.: Behavior of braced and anchored walls in soils overlying rock. J. Geotech. Geoenviron. Eng. **127**(3), 225–233 (2001)

Analysis of Dielectric Properties and Influencing Factors of Zn Contaminated Soil

Jiaqi Li[1,2], Xianggui Xiao[1,2], Jipeng Wang[1,2], Zonghui Liu[1,2](✉), and Kang Lin[1,2]

[1] School of Civil Engineering, Guangxi University, Nanning 530000, Guangxi, China
lzh8@gxu.edu.cn

[2] Guangxi Key Laboratory of Disaster Prevention and Engineering Safety, Guangxi University, Nanning 530000, Guangxi, China

Abstract. Studying on the influence of heavy mental contents and moisture contents on dielectric properties of soil and its internal mechanism, which is the basis of quantitative detection of soil pollution by ground penetrating radar (GPR). Taking zinc contaminated soils as an example, soil cakes are made by adding zinc salt solution of different mass concentration into silty clay. Then, the complex permittivity of soils in different wet densities, different moisture contents were measured by vector network analyzer. The laws of variation of complex permittivity were summarized and grey relational analysis was carried out. Finally, the mechanism of influence of various factors on dielectric properties of zinc contaminated soil was explained by dielectric theory. The results show that the real part of complex permittivity of zinc contaminated soil is most affected by moisture content, which increased with the increase of moisture content, the correlation between real part and moisture content was 0.93. The imaginary part of complex permittivity is most affected by zinc salt content, which increased with the increase of zinc salt content, the correlation between imaginary part and zinc salt content was 0.77.

1 Introduction

Soil is an important natural resource, crucial for survival and development of human society. The problem of soil pollution in China is increasingly serious with rapid development of economy. According to a report on the national general survey of soil contamination issued in 2014, the over standard rate of heavy mental of soil pollution in China is 16.1%. Among them, cadmium, nickel and other heavy metals are the most prominent inorganic pollution [1]. It is urgent to carry out the assessment and remediation of heavy metal contaminated soil. Ground Penetrating Radar (GPR) is used more and more frequently in detection of contaminated soil because of its advantages of non-destructive and high efficiency. Its basic principle is to infer the dielectric properties of soil according to the reflected wave characteristics of GPR, and the pollution degrees of contaminated soil could be understood. The dielectric property of soil refers to the ability of soil medium to store and lose electrostatic energy with the electric field, which is represented by the real part and imaginary part of complex permittivity, respectively. On the one hand, after cadmium, nickel and other heavy metal ions penetrate into the soil, it will change the phase composition and structure of soil [2], which are reflected in the real and imaginary parts of the complex permittivity. On the other hand, complex permittivity is a key

© The Author(s), under exclusive license to Springer Nature Switzerland AG 2021
S. Shu et al. (Eds.): GeoChina 2021, SUCI, pp. 105–114, 2021.
https://doi.org/10.1007/978-3-030-80142-7_8

parameter for quantitative detection of soil pollution by GPR, and its value is affected by saturation and salt content of soil. Therefore, it is necessary to master the relationship between complex permittivity and soil physical parameters to realize quantitative detection of contaminated soil.

In recent years, domestic and overseas scholars have done a lot of research on soil dielectric properties and achieved some results: Liu [3] have found that there is a relationship between soil dielectric properties and its physical parameters; Gao [4] has found that moisture content affects the sensitivity of geological radar to the content change of heavy metal pollutants in soil; Neng [5], Qiao [6] and Liu [7] studied the soil dielectric properties with different moisture content and chromium (CR) concentration, and the soil permittivity increased with the increase of moisture content and chromium concentration was found. Dhiware M D [8] studied the influence of soil physical and chemical properties on its dielectric properties; Orange A [9] found that the dry density of soil was directly proportional to the permittivity, and the effect of dry density on the dielectric properties was related to soil type; Kabir H [10] found that soil moisture was the main factor affecting its dielectric properties. Until now, part of the current studies still regard the complex permittivity as a constant, which not only does not consider the influence of frequency on soil dielectric properties, but also rarely carries out the mechanical analysis on the changes of soil dielectric properties from the dielectric point of view, and does not establish the relationship between the complex permittivity and soil moisture content, dry density and salt content. In this paper, zinc pollution was taken as an example, the complex permittivity of soil with different moisture contents, densities and zinc pollution degrees was measured by vector network analyzer and calculation software, and the change rules were summarized. Then, the correlation analysis was carried out by GRA. Finally, the test results were analyzed from the perspective of Dielectrics. The results can help to provide theoretical basis for the quantitative detection of heavy metal contaminated soil.

2 Test Method and Results

2.1 Test Method

The experiments were divided into three groups by using single factor analysis method, studying the factors of moisture content, wet density and zinc ion content. Soil samples with different zinc ion contents (0, 200, 400, 600, 800, 1000, 1200, 1400, 1600, 2000, 2500 and 3000 $mg \cdot kg^{-1}$), different wet densities (1.5, 1.6, 1.7, 1.8, 1.9 and 2.0 $g \cdot cm^{-3}$) and different moisture contents (14, 15, 16, 17, 18, 19, 20 and 21%) were prepared. The soil used in the experiment was taken from a deep foundation pit 5 m deep in Nanning City, and the original soil is silty clay. The elemental composition of the initial soil was calibrated by Semi-quantitative Spectrometric Analysis. According to the current national standard *Soil environmental quality agricultural land soil pollution risk control standard* (Trial) (GB 15618–2018), the risk screening value of zinc element was determined. The original soil can be considered as pollution-free soil. Then the original soil was dried, crushed and sieved. Calculating the required solution mass based on designed parameters, dissolving the prepared reagent in the corresponding deionized water, shaking and melting evenly, and spraying evenly into the soil sample through the spray pot. After the preparation of soil samples, the soil samples were packaged with plastic wrap and kept for 12 h to mix water, reagents and soil as much as possible to ensure the uniformity of soil samples. Finally, the soil was pressed into samples in ring knife that was 60 mm diameter and 20 mm high. The compaction process is shown in Fig. 1.

The complex permittivity of soil samples was measured by KEYSIGHT E5061B vector network analyzer. The frequency range was 100 ~ 3000 MHz, the sampling interval was 25 MHz, and the test environment was 20 ± 1 °C. The test method was open circuit reflection coaxial probe terminal. During the measurement, the samples are placed on the gravity scale, and the contact strength between coaxial probe and each sample is ensured to be consistent by observing the reading of the gravity scale. Vector network analyzer and coaxial open circuit terminal probe were shown in Fig. 2 and Fig. 3, respectively.

Fig. 1. Compaction of soil sample

Fig. 2. Vector network analyzers

Fig. 3. Coaxial terminal probe

2.2 Test Result

ε'_r and ε''_r stand for the real part and imaginary part of complex permittivity, respectively. Figure 4 shows the variation curves of complex permittivity of pollution-free soil with different moisture contents (Density is 2.0 g · cm^{-3}). It can be seen that both real and imaginary parts of soil complex permittivity increase with the increase of moisture content. In the low frequency range, both real and imaginary part decrease with the increase of frequency, and the decrease trend of imaginary part become more obvious when the moisture content become higher. In the high frequency range, the real part gradually tends to be stable while the imaginary part shows an increasing trend, and the increase trend become more obvious when the moisture content become higher.

Figure 5 shows the variation curves of complex permittivity of pollution-free soil with different wet densities (moisture content is 17%). It can be seen that both real and imaginary parts of soil complex permittivity increase with the increase of wet density. In the low frequency range, both real and imaginary part decrease with the increase of frequency. In the high frequency range, the real part tends to be stable while the imaginary part increases.

Figure 6 shows the variation curves of soil complex permittivity with different zinc pollution degrees (moisture content is 17%, density is 2.0 g·cm^{-3}). It can be seen that both real and imaginary parts of soil complex permittivity increase with the increase of zinc pollution. In the low frequency range, both real and imaginary part decrease with the increase of frequency. When the concentration of zinc ions become higher and the imaginary part become larger, the downward trend of imaginary part with frequency become more obvious. In the high frequency range, the real part gradually tends to be stable while the imaginary part shows an increasing trend. The increasing trend of imaginary part become more obvious when the zinc ion contents become lower.

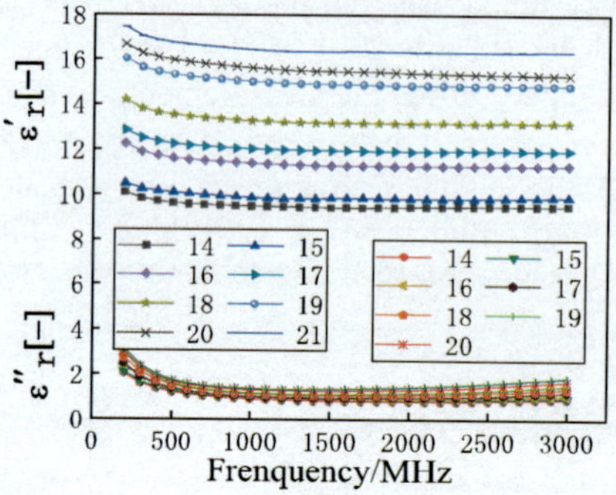

Fig. 4. Variation curves of complex permittivity of pollution-free soil with different moisture contents

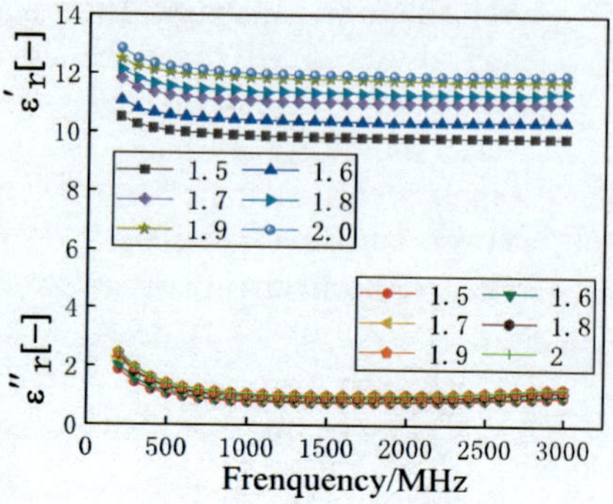

Fig. 5. Variation curves of complex permittivity of pollution-free soil with different wet densities

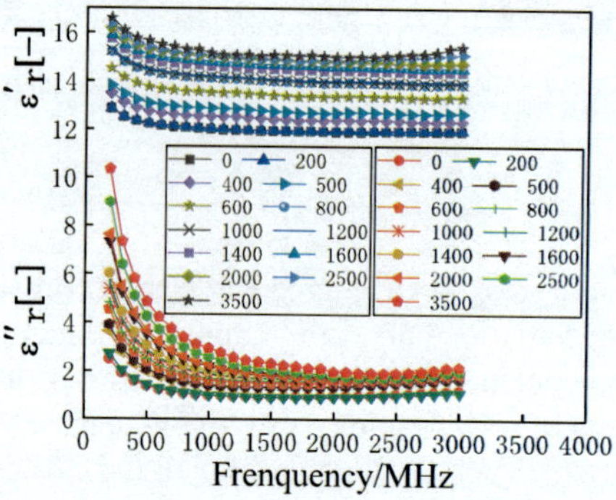

Fig. 6. Variation curves of soil complex permittivity with different degrees of zinc pollution

2.3 Grey Correlation Analysis

Correlation degree is a measure of correlation between different factors in one system and the value of that is high when the change trend of the two factors is consistent. Grey Relation Analysis is to analyze the influence degree of each factor on the main behavior of the system. The degree of correlation refers to the degree of geometric differences between curves. The initialization method is used for dimensionless processing of the experimental data before the analysis to reduce analysis error. The calculation process is as follows formula (1) ~ (2) [11].

$$\xi_{0i} = \frac{\Delta(\min) + \rho\Delta(\max)}{\Delta_{0i}(\kappa) + \rho\Delta(\max)} \tag{1}$$

$$r = \frac{1}{n}\sum_{i=1}^{n} \xi_i(\kappa) \tag{2}$$

Among them, ξ_{0i} = correlation coefficient, $\Delta(\max)$ = maximum difference of two levels, $\Delta(\min)$ = minimum difference of two levels, $\Delta_{0i}(K)$ = the absolute difference between each point on the x_i curve of each comparison sequence and each point on the reference sequence x_0 curve, ρ = distinguishing coefficient, the value is 0.5, r = correlation degree, k = different time or point, the value is 1,2,3…n.

Figure 7 shows the grey correlation coefficient curves of different factors and the real and imaginary parts of soil complex permittivity. The black, red, blue and pink lines represent the grey correlation coefficients of complex permittivity with moisture content, frequency, salt content and density, respectively. It can be seen from Fig. 7(a) that the black curve is above the other three curves, which means that the real part is affected by moisture content mostly and the correlation coefficients is 0.9285. Similarly, blue curve is above the other three curves in Fig. 7(b), which means the imaginary part is affected by salt content mostly and their correlation coefficients is 0.7688.

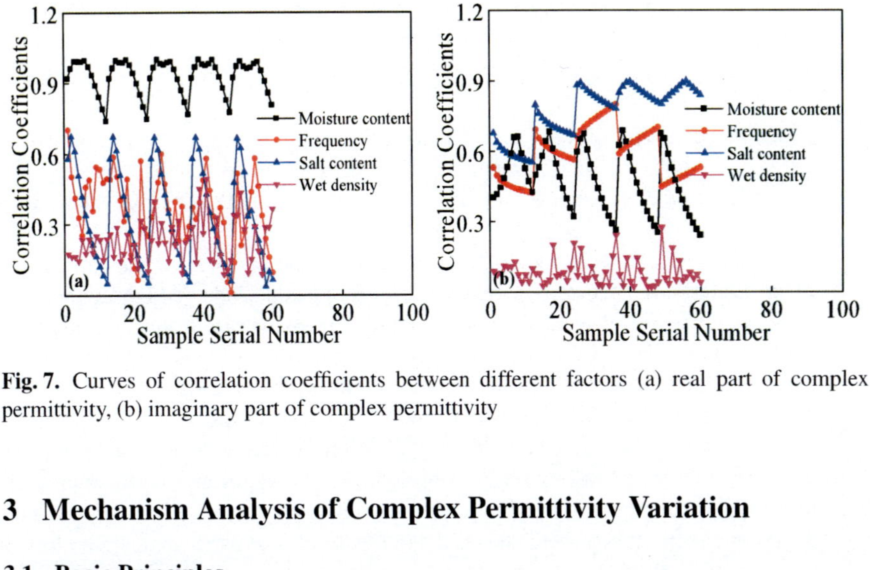

Fig. 7. Curves of correlation coefficients between different factors (a) real part of complex permittivity, (b) imaginary part of complex permittivity

3 Mechanism Analysis of Complex Permittivity Variation

3.1 Basic Principles

Permittivity is a physical quantity to characterizes the ability of substance to keep charge. The permittivity in constant electric field is a real number, while the permittivity in alternating electric field is a complex number, which is called complex permittivity. It is composed of the real part representing the ability to store and release electric energy instantaneously and the imaginary part representing the loss of power. The free charge in the medium begin to move, produce many electric dipoles arranged along the direction of external electric field and form a certain polarization intensity with an external alternating electric field, which called polarization of the medium. The polarization intensity is a physical quantity that characterizes the polarization degree of medium. According to the Debye relaxation relation:

$$\varepsilon_r' = \frac{\varepsilon_\infty + (\varepsilon_s - \varepsilon_\infty)}{(1 + \omega^2\tau^2)},\ \varepsilon_r'' = \frac{(\varepsilon_s - \varepsilon_\infty)\omega\tau}{(1 + \omega^2\tau^2)} \tag{3}$$

$$\tan\delta = \frac{\sigma}{\omega\varepsilon_e'} = \frac{\varepsilon_e''}{\varepsilon_e'} \tag{4}$$

Among them, ε_s = the permittivity in electrostatic field, ε_∞ = the permittivity in optical frequency, ω = the frequency of applied electric field. τ = relaxation time. The process of polarization can be divided into three stages: First, when the frequency is low, the electric field changes slowly so that the dielectric polarization can completely keep up with the change of external electric field, and the numerical value is equal to the static permittivity ε_s. Then, With the increase of frequency, the dielectric begins to fail to lose pace with the change of electric field, and the loss of dielectric begins to increase and reach the extreme value in a period of time. At last, when the frequency increases to a certain value, the polarization can not keep up with the change of external electric field completely because the electric field changes rapidly. Only instantaneous polarization occurs in medium, and ε_r' gradually reduces to optical frequency permittivity ε_∞.

3.2 Influence Mechanism of Frequency on Dielectric Properties

It can be seen from Fig. 4 to Fig. 6 that the real part of the complex permittivity decreases with the frequency at low frequency and tends to be stable at high frequency. According to formula 3–1, the real part is inversely proportional to the frequency, when the polarization of the medium is in second stage so that the phenomenon of polarization relaxation appears in the medium, which leads to the decrease of the energy obtained by the material molecules. The real part is thus reduced. Then, in high frequency range, the contaminated soil gradually tends to state of unpolarization, and the real part finally tends to be stable. On the other hand, the imaginary part of the complex permittivity decreases with frequency at low frequency and increases with frequency at high frequency. Because conductivity loss is the main dielectric loss in low frequency range. With the increase of frequency, the polarization degree, loss and the imaginary part decreases; When frequency range is high, the polarization loss is the main dielectric loss. With the increase of frequency, the polarization increases, the loss of medium and the imaginary part increases. The results in Fig. 6 show that the conductivity loss of free charge always dominates when zinc ions invade the soil because zinc ions have higher conductivity than water molecules.

3.3 Influence Mechanism of Different Factors on Dielectric Properties

Figure 8 shows the variation curves of soil complex permittivity with moisture content at certain wet density and zinc pollution degree in different frequencies. It can be seen that both the real part and the imaginary part are positively correlated with moisture content in different frequencies, and the increasing proportion of real part with moisture content is greater than that of imaginary part at the same frequency. The permittivity of soil as a whole is determined by permittivity of the three phase in soil. The relative permittivity of air is about 1 [12], while the value of relative permittivity of water at room temperature is 76.6. The increase of moisture content not only increases the content of water in soil, but also reduces the volume of air in soil, which increases the permittivity of soil. In addition, bond water is the main form of water in soil and the ion mobility is poor when the moisture content is low. With the increase of moisture content, the ion movement is enhanced and the polarization is intensified [13]. Therefore, the real part increases with the increase of moisture content. However, the intensification of ion motion also increases the probability of collision, resulting in the increase of conductivity loss, so the imaginary part also increases with the increase of moisture content.

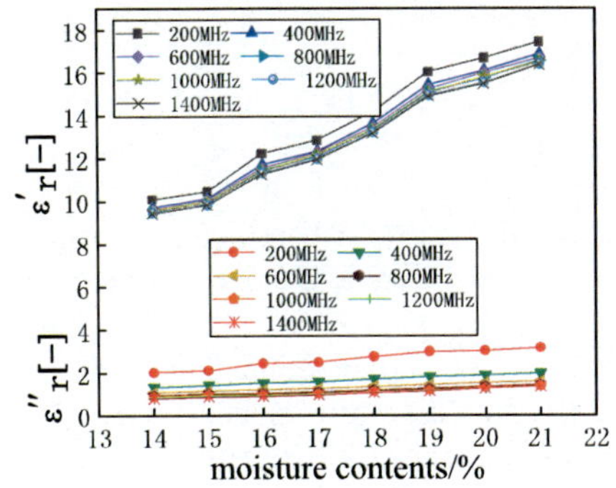

Fig. 8. Variation curves of complex permittivity of soil with zinc pollution with moisture contents

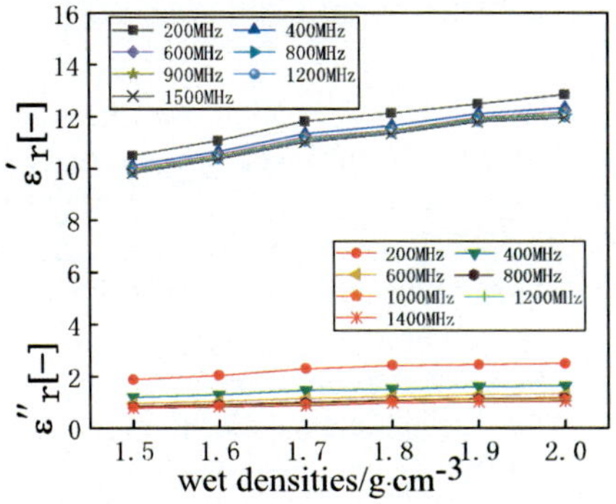

Fig. 9. Variation curves of complex permittivity of zinc contaminated soil with wet densities

Figure 9 shows the variation curves of soil complex permittivity with wet density at certain zinc pollution degree and moisture content in different frequencies. It can be seen that the real part and imaginary part are positively correlated with wet density in different frequencies.Similar to the effect of variation of moisture content, the increase of wet density not only reduces the volume of air in the soil, but also increases the mass of water and solid particles, which increases the permittivity of the whole soil. On the other hand, the increase of wet density means that particles in soil are arranged more closely, and the probability of collision increases when the medium moves with the external electric field, which intensifies the electrical conductivity loss. As a result, the imaginary part increases.

Figure 10 shows the variation curves of soil complex permittivity with zinc pollution degree at certain moisture content and wet density in different frequencies. It can be seen that the complex permittivity of soil is positively correlated with the degree of zinc pollution in different frequencies. At the same frequency, the proportion of imaginary part with the increase of moisture content is far more than that of the real part, and the change trend of imaginary part with the degree of zinc pollution is quite different due to different frequency: its increasing proportion decreases sharply with the increase of frequency. With the increase of zinc ion concentration, the amount of free charge

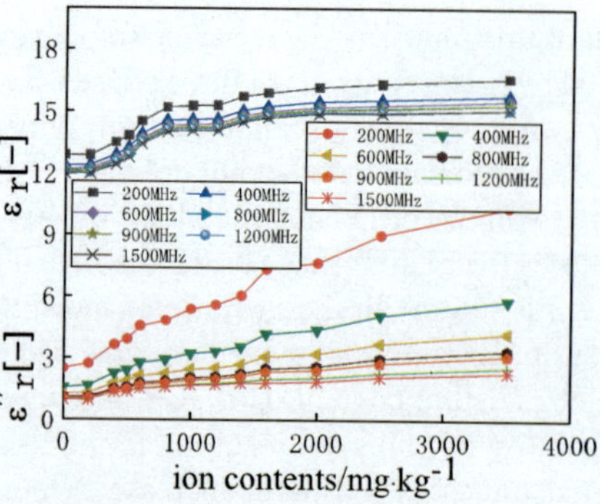

Fig. 10. Variation curve of complex permittivity of zinc contaminated soil with different ion contents

in soil increases. With the influence of external electric field, the polarization effect of medium is enhanced and the real part is increased. When the salt content is low, the dielectric loss caused by polarization effect of soil can be ignored, and the loss caused by soil polarization is large, which can not be ignored when the salt content is high [14]. Moreover, the increase of free charge in soil leads to the increase of ion collision probability, the imaginary part is affected by loss of polarization and conductivity, which makes that the increasing trend is far beyond the real part. In the previous study, it is found that the metal ion concentration has little effect on the electromagnetic wave velocity of GPR, which is inconsistent with the experimental results that the real part of complex permittivity increases with the increase of metal ion concentration. As far as the test results of this paper are concerned, without considering the factors such as experimental means and errors, it can be inferred that the moisture content of soil sample in this test is smaller than before. The variation trend of the real part with ion concentration depends on the sum of the enhancement of polarization effect of metal ions and the weakening of orientation polarization effect of water molecules. Therefore, the weakening of water molecular orientation polarization effect becomes not obvious when the moisture content becomes smaller, which leads to an obvious upward trend in the proportion of real part with the increase of zinc pollution degree. The variation of the real part of the complex permittivity with the concentration of metal ions needs to be further studied in the future.

4 Conclusion

- The complex permittivity of soil with different moisture contents, wet densities and zinc ion contents was measured by vector network analyzer, and the correlation coefficient of complex permittivity with moisture content, wet density, salt content of soil and external electric field frequency was obtained by Grey Relation Analysis. The results show that the real part of complex permittivity is affected by moisture content mostly, and the correlation degree was 0.9285. The imaginary part is affected by the salt content mostly, and the correlation degree is 0.7682. Variation trend of imaginary part with frequency is significantly different because of the differences of soil physical parameters.

- Based on the dielectric theory, the mechanism of complex dielectric properties of zinc contaminated soil was analyzed. The results show that the increase of moisture content, wet density and salt content can enhance the polarization effect of medium and increase the loss of medium in different degrees, resulting in the increase of the real and imaginary parts of complex permittivity. The increase of frequency weakens the polarization effect, leading to the decrease of real part of complex permittivity. The imaginary part decreases first and then increases because the dielectric loss is dominated by conductivity loss in low frequency range, while the dielectric loss is dominated by polarization loss in high frequency range.
- The value of permittivity obtained by vector network analyzer fluctuates to some extent because of the influence of test method, temperature and non-uniformity of soil sample. At present, the data can only qualitatively analyze the influence of various factors on the complex permittivity of zinc contaminated silty clay. More stable testing methods should be found in future study. A large number of tests should be carried out for more metal pollutants and soil types to establish quantitative relationship model.

Acknowledgments. This work was financially supported by the Natural Science Foundation for Young Scholars of Guangxi (Grant No. 2017GXNSFBA198199), the National Natural Science Foundation of China (Grant No. 51708136) and the Systematic Project of Guangxi Key Laboratory of Disaster Prevention and Engineering Safety (2019ZDX029).

References

1. Ma, J.F., Yan, J.Y.: The current situation of soil environmental pollution in China and its control measures. J. Environ. Dev. **2**, 48–49 (2020)
2. Song, Z.Z.: The study on micro-structure and engineering properties of heavy metal contaminated soil. J. Shanxi Architect. **42**(22), 91–92 (2016)
3. Liu, H.R., Neng, C.X., Wang, C., et al.: Review on detection of pollution content in the contaminated soil based on dielectric character. J. Prog. Geophys. **25**(6), 2184–2192 (2010)
4. Gao, Q., Zhao, G.Z., Li, Z.P., et al.: Analysis of vertical distribution characteristics and dielectric properties of heavy metal contaminants in soil. J. Geotech. Invest. Surveying. **46**(04), 68–74 (2018)
5. Neng, C.X., Liu, Y.Q., Liu, H.R., et al.: Experiment results of conduction, spectral induced polarization and dielectric characteristics for chrome-contaminated soil. J. Environ. Sci. **32**(3), 758–765 (2011)
6. Qiao, C.P., Zhao, G.Z., Liu, S.K., et al.: Study of the impact of heavy metals contamination on dielectric properties of soils. J. Irrig. Drainage **36**(8), 77–82 (2017)
7. Liu, S.K., Zhao, G.Z., Li, Z.P., et al.: Study on the influencing characteristics of soil dielectric properties with chromium (Cr) pollution. J. Geotech. Invest. Surveying **45**(4), 68–74 (2017)
8. Dhiware, M.D., Nahire, S.B., Deshmukh, S., et al.: Correlation between Physical and Chemical Properties of Soil and Dielectric Constant in Nasik Region, vol. 6, pp. 2325–2328 (2019)
9. Orangi, A., Narsilio, G.A., Wang, Y.H., Ryu, D.: Experimental investigation of dry density effects on dielectric properties of soil–water mixtures with different specific surface areas. Acta Geotech. **15**(5), 1153–1172 (2020). https://doi.org/10.1007/s11440-019-00805-x

10. Kabir, H., Khan, M.J., Brodie, G., et al.: Measurement and modelling of soil dielectric properties as a function of soil class and moisture content. J. Microw. Power Electromagn. Energ. Publ. Int. Microw. Power Inst. **2**, 1–16 (2020)
11. Zhang, X.J., Cheng, L.S., Li, C.L.: Application of grey correlation analysis method in analysis of influencing factors of production decline rate. J. Pet. Geol. Recovery Effi. **06**(11), 48–50 (2004)
12. Topp, G.C., Davis, J.L., Annan, A.P., et al.: Electromagnetic determination of soil water content: measurements in coaxial transmission lines. J. Water Resour. Res. **16**(3), 574–582 (1980)
13. Wensink, W.A.: Dielectric properties of wet soils in the frequency range 1–3000MHZ. J. Geophys. Prospect. **41**(6), 671–696 (1993)
14. Zhang, P.: Analysis to Effects of Main Factors on Dielectric Properties of Soils. Northwest A & F University (2013)

Author Index

© The Editor(s) (if applicable) and The Author(s), under exclusive license
to Springer Nature Switzerland AG 2021
S. Shu et al. (Eds.): GeoChina 2021, SUCI, p. 115, 2021.
https://doi.org/10.1007/978-3-030-80142-7